RAND NATIONAL DEFENSE RESEARCH INSTITUTE

An Assessment of Gaps in Business Acumen and Knowledge of Industry Within the Defense Acquisition Workforce

A Report Prepared for the U.S. Department of Defense in Compliance with Section 843(c) of the Fiscal Year 2018 National Defense Authorization Act

Laura Werber, John A. Ausink, Lindsay Daugherty, Brian Phillips, Felix Knutson, Ryan Haberman

Prepared for the Office of the Secretary of Defense

Approved for public release; distribution unlimited

For more information on this publication, visit www.rand.org/t/RR2825

Library of Congress Cataloging-in-Publication Data is available for this publication.
ISBN: 978-1-9774-0205-9

Published by the RAND Corporation, Santa Monica, Calif.
© Copyright 2019 RAND Corporation
RAND® is a registered trademark.

Support RAND

Make a tax-deductible charitable contribution at
www.rand.org/giving/contribute

www.rand.org

Preface

Section 843(c) of the National Defense Authorization Act of fiscal year 2018 directed the Under Secretary of Defense for Acquisition and Sustainment (USD[A&S]) to conduct an assessment of training for the acquisition workforce. The legislation indicated two key objectives of the assessment: (1) to determine the effectiveness of training and development resources offered by providers outside the U.S. Department of Defense (DoD) that were available to acquisition workforce (AWF) personnel and (2) to assess gaps in business acumen, knowledge of industry operations, and knowledge of industry motivation present within the AWF.

In April 2018, the USD(A&S) asked the RAND Corporation to carry out a study to accomplish the assessment. RAND researchers used a mixed-methods approach to conduct this assessment, including interviews with DoD and industry leaders, a review of AWF competency models and Defense Acquisition University course offerings, and literature and policy reviews. Findings correspond to the report elements specified in Section 843(c), most notably:

- the training and development options DoD was using at the time of the study to confer business acumen, knowledge of industry operations, and knowledge of industry motivation
- evidence of training gaps related to these three types of knowledge
- the potential role of training and development offered by external providers, such as industry firms and colleges and universities, in building business acumen and knowledge of industry within the AWF.

This research should be of interest to DoD personnel involved with AWF training and development and to congressional representatives and staff responsible for defense acquisition oversight. Some expertise about government civilian and military personnel management and defense acquisition is presumed in a reader. This research was sponsored by USD(A&S) and conducted within the Forces and Resources Policy Center of the RAND National Defense Research Institute, a federally funded research and development center sponsored by the Office of the Secretary of Defense, the Joint Staff, the Unified Combatant Commands, the Navy, the Marine Corps, the defense agencies, and the defense Intelligence Community. For more information on the RAND Forces and Resources Policy Center, see www.rand.org/nsrd/ndri/centers/frp or contact the director (contact information is provided on the web page).

Contents

Figures and Tables

Figures

Tables

Summary

More than 169,000 civilian and military personnel in the U.S. Department of Defense's (DoD's) acquisition workforce (AWF) "are responsible for identifying, developing, buying, and managing goods and services to support the military."[1] DoD is charged with developing a highly skilled AWF that is managed "in a manner that complements and reinforces the management of the defense acquisition system,"[2] and, since passage of the Defense Acquisition Workforce Improvement Act (DAWIA) in 1990, Congress has often expressed its interest in ensuring that this is accomplished effectively.

The fiscal year (FY) 2018 National Defense Authorization Act (NDAA) is one of the latest instances of this interest. Section 843 of the act directed the Under Secretary of Defense for Acquisition and Sustainment (USD[A&S]) to evaluate business-related training for the AWF, focusing on potential knowledge gaps related to business acumen, industry operations, or industry motivation. Opportunities to use training provided by industry or other external sources to close these gaps were to be part of the evaluation.

The act specified four specific areas to include in the assessment:

---

[1] M. Schwartz, K. A. Francis, and C. V. O'Connor, *The Department of Defense Acquisition Workforce: Background, Analysis, and Questions for Congress*, Washington, D.C., Congressional Research Service, CRS Report R44758, July 29, 2016. Schwartz references a RAND report for this definition: Susan M. Gates, Edward G. Keating, Adria D. Jewell, Lindsay Daugherty, Bryan Tysinger, Albert A. Robbert, and Ralph Masi, *The Defense Acquisition Workforce: An Analysis of Personnel Trends Relevant to Policy, 1993–2006*, Santa Monica, Calif.: RAND Corporation, TR-572-OSD, 2008, p. 2.

Schwartz suggests a more detailed definition as follows: "Generally, the acquisition workforce consists of uniformed and civilian personnel who are either

- in positions designated as part of the acquisition workforce under the Defense Acquisition Workforce Improvement Act (10 U.S.C. §1721);
- in positions designated as part of the acquisition workforce by the heads of the relevant military component, pursuant to DOD Instruction 5000.66; or
- temporary members of the acquisition workforce or personnel who contribute significantly to the process, as defined in the Defense Acquisition Workforce Development Fund (10 U.S.C. §1705)."

[2] 10 U.S.C. paragraph 1701a, *Management for Acquisition Workforce Excellence*.

1. current sources of training and development (T&D) opportunities related to business acumen, knowledge of industry operations, and knowledge of industry motivation
2. knowledge gaps related to business acumen, industry operations, and industry motivation for each acquisition position
3. plans to address those gaps for each acquisition position
4. consideration of the role that organizations outside of the Defense Acquisition University (DAU) could play in addressing gaps.

To address these areas, RAND researchers examined

- how AWF learning requirements are developed and communicated and, in particular, how the need for the types of knowledge mentioned in Section 843 is documented
- gaps in the areas of knowledge of concern to Congress
- resources available from DoD and non-DoD sources to confer this knowledge and thereby avoid or close gaps
- approaches to assessing the provision of T&D that might help DoD decide the appropriate resources to use in providing the training of interest to Congress.

Approach

To meet the congressionally mandated timeline, RAND researchers had six months to complete this study. This short time period had implications for the methods available to the study team. For example, it was infeasible to interview or survey members of the AWF. To overcome this and other limitations, the study team relied on a mixed-methods approach to the assessment that included interviews and targeted discussions with experts familiar with the AWF, a review of competencies expected of AWF personnel and of DAU course offerings, and literature and policy reviews.

Interviews with subject-matter experts (SMEs) inside and outside DoD were a primary source of data for this study. We conducted 44 semi-structured interviews with DoD senior leaders and SMEs and with external stakeholders. The 26 DoD interviews included key personnel tasked with AWF career-management responsibilities, including directors for acquisition career management (DACMs) for the three military departments and DoD's Fourth Estate, functional leaders (or their designees) for all of the AWF career fields, DoD Human Capital Initiatives (HCI) leadership, and directors (or acting directors) from all five of DAU's centers. Participants in the 18 external stakeholder interviews included representatives of professional associations, private-sector organizations with industry rotation programs and/or extensive in-house corporate universities, and universities that were providing customized courses for

AWF personnel at the time of this study. We also spoke with members of the FY 2016 NDAA Section 809 panel, a congressionally mandated panel tasked to identify ways to improve the defense acquisition process.

Our second primary data collection effort consisted of targeted discussions with DoD SMEs intended to help us to learn more about DoD T&D efforts in use at the time of the study. These differed from the aforementioned interviews in that, rather than repeating similar questions across sessions, we asked a limited, unique set of questions pertinent to the SME's background.

The study results were also informed by a wide array of secondary data sources. Among these sources were those pertaining to the competency management framework DoD uses for its strategic human capital planning. This framework uses competency models to align mission outcomes with expectations for employee behaviors, where a *competency* is defined as "an observable, measurable pattern of knowledge, abilities, skills, and other characteristics that individuals need to perform work roles or occupational functions successfully."[3] We reviewed DoD instructions (DoDIs) and documents that govern the development and use of competency models for AWF career fields. We examined the most up-to-date career-field competency models available, as well as lists of competencies provided in Acquisition Workforce Qualification Initiative e-workbooks (which were available online) to determine the extent to which they included competency elements corresponding to Section 843–related knowledge. We also examined DAU course requirements for career-field DAWIA certifications to see how needs for these types of knowledge were reflected in required courses.

Other secondary works that we reviewed included the following:

- DAU publications, such as its catalog, course descriptions, and directives
- documentation related to degree-granting opportunities, executive training options, and individual courses provided by other DoD institutions, such as the Naval Postgraduate School (NPS) and the Air Force Institute of Technology (AFIT)
- online and published information on degree-granting programs, executive training programs, and training-with-industry programs available to AWF members via external (non-DoD) providers
- publications about the AWF pertaining to its competencies, T&D, or knowledge gaps
- literature around two specific topics of interest: evidence on best practices for the design and delivery of T&D and evidence on best practices for the evaluation of T&D
- published company-specific examples of T&D practices.

---

[3]   DoD, *DoD Civilian Personnel Management System: Civilian Strategic Human Capital Planning (SHCP)*, DoDI 1400.25, Vol. 250, Enclosure 5, June 7, 2016a.

Overall, we sought, to the greatest extent possible, to base our findings on multiple data sources. For example, our assessment of knowledge gaps was informed by our competency model analysis, interviews, and review of related publications. Similarly, our examination of potential T&D-related strengths and weaknesses drew again from our interviews, as well as from published information on external T&D options, literature on best practices for the design and delivery of T&D and its evaluation, and company-specific practices.

Findings

The Lack of Standardized Definitions and Competency Model Formats Obscures the Need for Knowledge Related to Business Acumen, Industry Operations, and Industry Motivation

We examined the application of the competency management framework to the need for all three types of Section 843–related knowledge and found the following:

- **There are no standardized definitions of the terms highlighted in Section 843.** There is a definition of business acumen in executive core qualifications (ECQs) for members of the government's Senior Executive Service, but this is neither specific to the AWF nor applicable to all its members, and variations of the definition exist in other AWF documents. We were unable to locate definitions for industry operations or industry motivation. As a result, we developed working definitions based on those offered by our DoD interviewees, which were subsequently approved for project use by our research sponsor. Those definitions are provided in Figure S.1. The lack of standardized definitions complicated our search of career field–level competency models for the need expressed for these types of knowledge and made it more difficult to determine how needs for these types of knowledge were satisfied by DAU learning assets.
- **Competency models are not developed in standardized formats.** In addition to the difficulties presented by a lack of official definitions for the areas of knowledge highlighted in Section 843, determining the need for these types of knowledge was hampered by the fact that AWF career-field competency model formats varied. Some of them introduce terms not used in the overall DoD guidance for competency models. One described desired levels of proficiency in terms of the five competency levels of the DoD guidance, while another described them in terms of DAWIA certification levels (I, II, or III). Most made no distinction among proficiency levels.
- **Competency models are developed with limited coordination across career fields, and there is no common structure to map competencies to career progression.** AWF competency models for different career fields are developed by

Figure S.1
Definitions of Business Acumen, Industry Operations, and Industry Motivation

Business acumen	Industry operations	Industry motivation
In addition to the ability to manage human, financial, and information resources strategically (OPM definition), business acumen is an understanding of industry behavior and trends that enables one to shape smart business decisions for the government.	This includes plans and procedures used within an industry to provide a product or service. The need for knowledge of specific practices may vary, depending on an employee's contribution to the acquisition mission. Some industry operations may be business oriented, while others may be at the confluence of business and technical knowledge—i.e., "techno-business" (e.g., milestone reviews).	This includes the range of considerations and motivations that factor into the decisionmaking of organizations in industry, including profit and revenue, market share, management and employee incentives, shareholder considerations, perspectives on risk, and the need to maintain position in a competitive environment. The relative weights of these factors may vary by industry and over time.

NOTE: OPM = Office of Personnel Management.

career-field functional leaders and functional integrated product teams (FIPTs). With the exception of Program Management, which completed a competency assessment and revision in 2015, all functional leaders are working with the HCI organization and the Defense Civilian Personnel Advisory Service (DCPAS) to revise their models using the Defense Competency Assessment Tool (DCAT).[4] However, the DCAT effort focuses on technical competencies, excluding many knowledge elements that are covered by the Section 843 language. Acquisition Workforce Qualification Initiative (AWQI) "e-workbooks" link competencies to mission-required products and their associated tasks to allow an employee to capture their demonstrated experience, but they do not include competencies that are deemed common expectations for employees outside the AWF. Finally, some services have developed their own competency models to address service-specific needs.

- **Competency models are developed and revised differently across the career fields.** Despite the formal guidance provided in DoD instructions that seeks to standardize the career-field competency models, the models that we obtained had inconsistencies in their structures, levels of detail, and how recently they appear to have been updated.

---

[4]   The Program Management competency model revision satisfied all DCPAS requirements, so the Program Management career field will participate only in the validation phase of the new effort, which includes convening a SME panel to discuss data that has been run through the DCAT process. (e-mail communications from HCI, January 23, 2019, and January 30, 2019).

Recognizing these constraints, our analysis does not attempt to determine conclusively which acquisition career fields have a need for business acumen, knowledge of industry operations, or knowledge of industry motivation. However, using our definitions of these three types of knowledge, we do show that AWF competency models and DAWIA certification requirements are fairly consistent in the expression of the relative need of different career fields for these types of knowledge. Our analysis of the career field–level competency models and DAU courses required for DAWIA certification shows that the need for these three types of knowledge, as indicated by these sources, exists for all career fields. This analysis, coupled with our interview results, further indicates that the relative need for the knowledge types varies across career fields.

Knowledge Gaps in Business Acumen, Industry Operations, and Industry Motivation Exist, but the Lack of Requirements and Desired Proficiencies Precludes an Estimation of Their Extent

Because a baseline of requirements and desired proficiency levels for the types of knowledge in Section 843 does not exist, it was challenging to gauge the extent to which they were lacking in the AWF workforce overall—much less on a career field or position basis. Consequently, we relied on our interviews with DoD leaders and external stakeholders to identify perceived gaps in business acumen, knowledge of industry operations, and knowledge of industry motivation, and we sought to corroborate interview findings with gap-related studies as much as possible.

Based on those sources, we conclude that gaps related to business acumen, knowledge of industry operations, and knowledge of industry motivation are present within the AWF to an indeterminate extent. Specific aspects of business acumen mentioned by interviewees include risk management and earned value management (EVM). EVM was identified in studies and interviews as a competency needed in multiple career fields. Aspects of industry operations–related knowledge deemed as both important and deficient include financial aspects, supply chain management, small business, agile development, and cybersecurity. The strongest evidence from the interviews was related to industry financial practices, such as financial management–related operations, corporate financial documents, and industry accounting. Knowledge of industry motivation was also perceived as critical yet lacking within the AWF, and interviewees emphasized elements, such as incentives, that drive corporate decisionmaking and actions and the influence of executive compensation structure. Gaps related to industry knowledge were also perceived as having an influence on other important types of knowledge, skills, and abilities important to the AWF: negotiation, developing and understanding requirements, and cost and price analysis.

A Variety of Internal and External T&D Assets Are Used, but Training Gaps Related to Section 843 Knowledge Were Difficult to Determine

Those tasked with AWF career-management responsibilities employ a wide array of internal (DoD-based) and external (non-DoD) resources to provide the three types of knowledge addressed by Section 843. Internal resources include T&D activities offered by the following:

- DAU, which provides a variety of learning assets (e.g., classroom and online courses, web-based references, workshops, and mission assistance teams)
- Air Force Institute of Technology courses and academic degree programs
- Naval Postgraduate School courses and academic degree programs
- the Eisenhower School for National Security and Research Strategy
- training resources provided by individual services.

DoD also routinely uses internal rotational assignments as a form of career development, and interviewees cited on-the-job training in discussions of DoD-provided T&D.

External resources include

- public and private colleges and universities that offer certificate and degree programs and executive training
- defense and non-defense corporations that conduct training through in-house corporate universities or academies
- professional associations that create training and examinations for professional certifications
- commercial vendors that provide DAU-equivalent courses
- industries that participate in rotational programs—e.g., service-specific training-with-industry programs and the Secretary of Defense Executive Fellowship Program.

All of these resources are intended to provide Section 843–related knowledge to varying degrees, but their capacity to do so appears to vary. For example, access to online DAU courses is virtually unlimited, but DAU's ACQ 315 course, "Understanding Industry," which was developed in response to concerns about AWF business acumen, appears to have an annual capacity of around 1,300. Access to external T&D opportunities also varied, from unlimited access to commercial courses and certificate programs to a very small number of slots in specialized rotation, executive education, and degree programs. In an example of the latter, for the programs we learned about, annual participants in rotations with industry ranged from three to 50.

Overall, DoD appears to be utilizing the range of approaches that industry offers to train and develop personnel; no training gaps with respect to the *types* of T&D options used are apparent. Yet our interviews featured ample discussions of the potential benefits of greater use of non-DoD T&D options—especially industry rotations—suggesting a perception that external T&D offerings are insufficient. Interviewees also agreed that the incorporation of industry resources and participants in internal T&D was valuable and that more could be done in this area.

Data limitations prevented us from making conclusive findings about the need for more external T&D. While we found limited capacity for external T&D offerings, we could not determine whether this capacity was adequate because of a lack of data in two areas: (1) data on which AWF personnel need the types of knowledge specified in Section 843 and (2) data on which AWF personnel have participated in internal and external T&D to address these knowledge requirements.

A full description of training gaps requires a systematic assessment of the entire portfolio of internal and external offerings along common criteria (e.g., cost, quality, effectiveness), and it was outside the scope of the study to conduct this broader analysis of the DoD training portfolio. However, our evidence suggests that external T&D may fill gaps in particular areas of the portfolio, such as offering opportunities for experiential learning and training alongside experts in industry. On the other hand, interviewees also identified substantial barriers to the use of industry providers to address T&D needs, including limitations on backfilling positions for civilian participants in industry rotations and limited access to funding for external T&D.

Monitoring the Effectiveness of T&D Related to Section 843 Knowledge Is Limited
As expressed in Section 843, Congress is interested in assessing "[t]he effectiveness of industry certifications, other industry training programs, including fellowships, and training and education programs at educational institutions outside of the Defense Acquisition University available to defense acquisition workforce personnel." To address this interest, we reviewed best practices in the evaluation of corporate T&D and compared DoD's current practices with industry best practices.

In our review of best practices for the evaluation of corporate T&D, we found that the Kirkpatrick Model was the most-cited approach in the literature and was referred to as the industry standard.[5] The approach includes four levels of evaluation: reaction (level 1), learning (level 2), behavior (level 3), and results (level 4). Some researchers also include a fifth level, return on investment (ROI). The Office of Personnel Management (OPM) used the Kirkpatrick Model as the framework for its *Training Evalu-*

---

[5] W. Arthur, Jr., W. Bennett, Jr., P. S. Edens, and S. T. Bell, "Effectiveness of Training in Organizations: A Meta-Analysis of Design and Evaluation Features," *Journal of Applied Psychology*, Vol. 88, No. 2, 2003, p. 234; Executive Development Associates, *Trends in Executive Development 2014: A Benchmark Report*, London: Pearson, undated; D. Wentworth, H. Tompson, M. Vickers, A. Paradise, and M. Czarnowsky, *The Value of Evaluation: Making Training Evaluations More Effective*, Alexandria, Va.: American Society for Training & Development, 2009.

ation Field Guide, and several large-scale surveys of corporate T&D evaluation were structured around the Kirkpatrick Model.[6]

Organizations most frequently evaluate T&D at the reaction and learning levels, with considerably fewer conducting evaluations at higher levels. We found a plethora of measures and data collection methods for each level, with surveys of employee or student satisfaction and trainer impressions prevailing at level 1. Additional evaluation strategies at higher levels include interviews, focus groups, knowledge tests, skill observations, customer/client assessments, peer evaluations, supervisor feedback, and reviews of key business metrics. Our review showed that overall, there is no one best way to evaluate T&D, especially at the higher levels of the Kirkpatrick Model. Impacts of interest vary by sector as well as by organization, and as they differ, so too may the right mix of measures and data collection techniques.

Turning our attention to DoD, we noted that DAU strives to evaluate its T&D offerings at the first four levels of the Kirkpatrick Model. However, our interviews suggested that, similar to many organizations, DoD efforts are primarily at levels 1 and 2. We found that DoD's in-use evaluation practices are relatively limited, particularly for the external T&D activities that Congress highlighted. Efforts to gauge the effectiveness of industry rotations are highly variable across fellowship programs, for example, and, in general, evaluations that are conducted are based on the student's or participant's own perception of the training and development's usefulness, how it affected his or her proficiency level, and how it affected his or her work performance.

We found few efforts to provide objective measures of the effectiveness of T&D, either in terms of demonstrated impact on the performance of individuals in their jobs or in terms of ROI. One of the stronger examples we found was for DAU 300- and 400-level courses. Specifically, 120 days after course completion, supervisors of graduates of these courses are asked to gauge how the course may have influenced the course graduates' work performance (Kirkpatrick Model level 4). Limited evidence of effectiveness makes it difficult to determine whether the "right" mix of T&D is in use and desired knowledge gains are attained, and this problem is exacerbated by the limited collection of data that tracks AWF participation in external T&D.

Recommendations

DoD is attempting to address gaps in the areas of business acumen, knowledge of industry operations, and knowledge of industry motivation, but it is difficult to determine the extent of these gaps, and without efforts to better estimate needs and effectiveness, gaps may persist. Challenges to such endeavors include

---

[6]   OPM, *Training Evaluation Field Guide: Demonstrating the Value of Training at Every Level,* Washington, D.C., 2011; Association for Talent Development, *Evaluating Learning: Getting to Measurements that Matter,* Alexandria, Va., 2016a.

- lack of defined knowledge requirements in section 843 areas
- lack of a formal gap assessment on section 843–related knowledge
- limited tracking of participation in external T&D
- limited efforts to assess T&D offerings as a portfolio
- limited data on the effectiveness of T&D.

The following recommendations will help DoD address these AWF-related challenges. We have grouped them into three categories: process-focused recommendations for DoD, external T&D–focused recommendations for DoD, and recommendations for Congress.[7]

Process-Focused Recommendations for DoD

Clarify the Nature and Extent of Needs for Business Acumen, Knowledge of Industry Operations, and Knowledge of Industry Motivation

As an important first step to determining learning objectives and the correct mix of training and development, DoD should clarify the nature and extent of needs for Section 843–related knowledge for each career field. This includes engaging Congress to develop agreed-upon definitions of business acumen, industry operations, and industry motivation, which DoD can then use to inform its competency models and other sources of knowledge requirements. It also involves considering the degree to which members of each career field need this knowledge in order to perform the tasks required of them, given their role in the acquisition mission.

Improve Approaches to Competency Assessments and Models

Functional leaders, perhaps through the Workforce Management Group (WMG), should coordinate the development of a standard format for competency models that conforms to the structure defined in DoDI 1400.25, Vol. 25, and DoDI 5000.66 and includes proficiency standards in line with the requirements in those instructions. It is also important that the functional leaders consider appropriate ways to incorporate industry standards and perspectives into these models and ensure that they are reviewed and updated at regular intervals.

Improve Approach to Knowledge Gap Assessments

DoD should develop and implement a rigorous approach to measuring proficiency in business acumen, knowledge of industry operations, and knowledge of industry motivation—ideally one that goes beyond relying on AWF professionals' self-reported proficiency and uses standard methods and measures across career fields so that results can be rolled up to the full AWF level.

---

[7]   We describe the qualitative benefits of these recommendations, but it was beyond the scope of this project to determine the costs of those recommendations that might require additional resources.

Improve Coordination of Internal and External T&D as a Single Portfolio of Offerings

With better understanding of various T&D options facilitated in part by a centralized, DoD enterprise–level repository of such opportunities, AWF career managers and personnel supervisors can make informed decisions about what training resources are appropriate for the career plans of those they supervise, and this repository may empower individual professionals to do more to chart their own career development courses. Improved coordination is also important to ensure that DoD makes optimal use of external T&D offerings—namely, to fill gaps in internal options rather than developing T&D activities that may be redundant or fail to address critical knowledge requirements.

Improve Tracking of Participation in T&D Activities that Confer Business Acumen, Knowledge of Industry Operations, and Knowledge of Industry Motivation

To identify training gaps, it is essential to understand which AWF personnel have received various types of T&D, including participation in external T&D. For example, the ability to track business degrees held by defense acquisition personnel, their prior experience working with industry, and other completed T&D activities in DoD's personnel management systems would make it easier to determine which AWF personnel are most lacking in exposure to Section 843–related knowledge and thus should have priority to participate in capacity-limited T&D activities. Individual-level tracking of T&D completed while in the AWF could also inform the evaluation of DoD's T&D offerings. Improvements to tracking may require significant information technology investments, as well as incentives for acquisition personnel to log their T&D experiences, particularly those obtained before joining the AWF.

Improve Evaluation of T&D

Understanding the needs that T&D is addressing is an important first step in identifying measures to track how effective various T&D offerings are in meeting those needs. Armed with this understanding, those tasked with AWF career T&D responsibilities should endeavor to track all T&D participation at the individual level, extend internal evaluation practices to external training, track career outcomes of T&D participants, and consider more-rigorous evaluations for costly programs.

External T&D–Focused Recommendations for DoD

Clarify and Enforce Reporting Requirements for Fellowships and Industry Rotations

DoDI 1322.06 directs military departments to provide an annual report to the Under Secretary of Defense for Personnel and Readiness (USD[P&R]) on their fellowships, internships, scholarships, training-with-industry programs, and grants to determine how the cost-effectiveness of each of these T&D options compares against others. However, the departments were not consistently meeting those reporting requirements at the time of this study. In addition, the DoDI does not specify how T&D effectiveness should be assessed and reported. USD(P&R) should address these problems with guidance on how the return on investment of these programs should be determined, to include specifying how indirect and direct costs should be calculated, as well as the standards and methods to use in evaluating fellowships and other T&D.

Further Assess the Need for Government–Industry "Co-Education"

Interviewees commented on the need for more government–industry "co-education" in two areas: industry rotations and use of industry resources (i.e., participants, presenters, standards) in internal T&D. Adopting the process recommendations described previously will provide DoD with the tools it needs to more conclusively determine whether there are training gaps that require more of these industry T&D resources. If there are gaps, ongoing efforts may help to fill some of them. For example, the Public-Private Talent Exchange authority is developing new industry rotation opportunities. The Deputy Secretary of Defense is encouraging "frequent, fair, even and transparent dialogue" with industry.[8] And interviewees described a range of ongoing efforts within DAU to incorporate industry resources into training.

If further assessment by DoD confirms a need for additional government–industry co-education, there are many possibilities for increasing use of these resources. For example, DAU could expand its use of industry experience as a type of criterion for evaluation of faculty position candidates and book more guest instructors and guest speakers from industry. An increase in the number of industry students in DAU in-residence classes could be a way to expose more AWF professionals to industry perspectives and vice versa. Looking to new companies as sources for rotations and corporate training could help to expand capacity. All of these efforts could result in more DAU students being exposed to industry perspectives, possibly at an earlier point in their educations.

Recommendations for Congress

We could not make a conclusive determination that DoD needs to use more industry-based training and external educational providers to address gaps related to business acumen, knowledge of industry operations, or knowledge of industry motivation. However, if Congress aims to incentivize use of external T&D providers, there are several policy options for addressing key barriers to the use of external T&D.

Relax Legislative Restrictions on Backfilling Positions When Personnel Participate in Industry Rotations

Industry rotations for civilian personnel do not permit DoD to backfill positions that are temporarily vacated by participants in the programs, and this was cited as a barrier to participating in the programs. If Congress wants to incentivize use of external T&D providers, it should consider accepting HCI's recommendation that the FY 2020

---

8   P. M. Shanahan, Deputy Secretary of Defense, "Engaging with Industry," memorandum, March 2, 2018a.

NDAA include an amendment to the Public-Private Talent Exchange authority established in Section 1104 of the FY 2017 NDAA to remove the backfill prohibition or, at a minimum, to allow waivers to this restriction in certain circumstances.

Promote Greater Use of the Defense Acquisition Workforce Development Fund

Concerns about the costs associated with external T&D and funding stability were cited as barriers to use of industry-based T&D and external educational providers. The Defense Acquisition Workforce Development Fund (DAWDF) can be tapped in many ways to build business acumen, knowledge of industry operations, and knowledge of industry motivation, and it increases the flexibility for DoD to consider a range of external options as part of its T&D portfolio. We recommend that Congress protect current funding levels and potentially raise them if DoD demonstrates that it can consistently execute the funds fully. Congress should continue to support DAWDF's various applications, which include both assistance for individual AWF members and component and DoD-wide T&D-related efforts, and it should encourage more expansive use of DAWDF to provide investment in the data infrastructure needed to track T&D assignments and accomplish meaningful evaluations of T&D effectiveness.

Give Actions Taken to Address AWF Knowledge Gaps Sufficient Time to Have an Effect

Effective solutions to closing knowledge gaps in the AWF—even those readily apparent—may take time to create and execute. In addition, the impact of an avoided or closed gap may not be immediately clear because some mission-related outcomes are long term. That stated, deadlines to ensure that improvements are made in a timely manner and that impacts are evaluated as soon as feasible may promote ongoing progress in meeting Section 843–related knowledge requirements. Intended outcomes may take years rather than months to be realized fully, but monitoring output-based measures (e.g., participation in industry rotations or attendance at industry-offered courses) on a regular basis could help to indicate future gains and inform evaluations of T&D effectiveness.

Conclusion

With neither standard definitions for knowledge related to business acumen, industry operations, and industry motivation nor estimates of the necessary proficiency in each, it was difficult to address Congress's questions about gaps in those areas and the best mix of T&D to close them. Nonetheless, our research showed that DoD uses a wide variety of internal and external T&D resources to avoid those gaps and takes targeted steps to close gaps once they become apparent. Clarifying definitions, standardizing competency models, and improving approaches to gap assessment will enable DoD to better determine which career fields have a need for access to more T&D opportunities. Managing internal and external T&D resources as a portfolio and improving the evaluation of their ability to confer Section 843–related knowledge will enable better decisions about the best use of those resources for different AWF populations. Congress has helped increase the opportunities for AWF personnel to learn from industry through means such as DAWDF and the new Public-Private Talent Exchange, but it can do more to enable DoD to take advantage of external training opportunities. Taken together, these efforts will help DoD to develop a highly skilled AWF.

Acknowledgments

We appreciate the research sponsorship of Kenneth Spiro, Chief of Staff, Office of the Assistant Secretary of Defense for Acquisition (ASD[A]). He also served as our action officer. We also benefited from the support provided by John Christian, ASD(A), and Richard Hoeferkamp, ASD(A) and Defense Acquisition University (DAU).

RAND researchers spoke with many leaders and subject-matter experts whose time, insights, and participation were of great value to this study. We acknowledge and thank the following participants from the U.S. Department of Defense (DoD):

Air Force
- Steven Clark, Chief, Air Force Industrial Liaison Office (SAF/AQXE)
- David Slade, Air Force Director for Acquisition Career Management (DACM)
- Michelle Trigg, Air Force Deputy DACM

Army
- James C. Dalton, Facilities Engineering Functional Leader
- Joan Sable, Division Chief, Army Acquisition Workforce Human Capital Initiatives (HCI), Army DACM Office
- Craig Spisak, Army DACM

Navy
- William Mark Deskins, Navy DACM
- Terry Emmert, former Life Cycle Logistics Functional Leader

Fourth Estate
- Anita Bales, Auditing Functional Leader
- Scott Bauer, Fourth Estate DACM
- Kevin M. Fahey, Program Management Functional Leader
- James Galvin, Acting Director, DoD Office of Small Business Programs
- Robert Gold, Executive Secretary, Engineering and Production, Quality, and Manufacturing Functional Integrated Product Teams (FIPTs); Engineering and

Production, Quality, and Manufacturing Functional Leader Interview Representative
- Brian Hall, Test and Evaluation Functional Leader
- Thomas Hickok, Executive Secretary, Information Technology FIPT
- Dale Ormond, Science and Technology Management Functional Leader
- Philip Rodgers, Business-Cost Estimating and Business Financial Management Deputy Functional Leader
- Bobbie Sanders, Information Technology Functional Leader
- Kenneth Spiro, Executive Secretary, Program Management FIPT
- Nancy Spruill, Business-Cost Estimating and Business Financial Management Functional Leader
- Jill Stiglich, Contracting, Purchasing, and Industrial Contract and Property Management Functional Leader Interview Representative
- René Thomas-Rizzo, Director, HCI, Office of the Under Secretary of Defense for Acquisition and Sustainment (A&S)

DAU
- Tom Davis, Industry Chair
- Sharon Jackson, Director, Business Center
- Bill Kobren, Director, Logistics & Sustainment Center
- Bill Parker, Director, Foundational Learning Directorate
- Randy Pilling, Director, Acquisition Management Center
- David Pearson, Director, Engineering & Technology Center
- Ray Ward, Acting Director, Contracting Center

We also are grateful for the support of our industry-based interview participants, whom we have not identified here for various reasons, including interviewee request and the need to avoid implied endorsement.

We thank the DoD subject-matter experts with whom we engaged in targeted discussions via telephone or email correspondence about specific aspects of AWF career management. They include Mark Camporini, DAU; Capt Katherine B. Hansen, Air Force Institute of Technology; Veronica Passarelli, Defense Civilian Personnel Advisory Service; Susan Pollack, Office of the Under Secretary of Defense for A&S; Jeb Ramsey, DAU; Barbara Smith, DAU; and Aissa Tovar, HCI, Office of the Under Secretary of Defense for A&S.

Our RAND colleagues John Winkler and Lisa Harrington provided guidance as the director and associate director of the RAND National Defense Research Institute (NDRI) Forces and Resources Policy Center. We received constructive reviews of an earlier version of this report from Mark Doboga, Charles Goldman, and Rene Rendon. Craig Bond and Sarah Meadows helped us navigate the RAND NDRI quality assurance process.

We thank them all, but we retain full responsibility for the objectivity, accuracy, and analytic integrity of the work presented in this report.

Abbreviations

A&S	Acquisition and Sustainment
ACE	American Council on Education
AFIT	Air Force Institute of Technology
ARP	Acquisition Research Program
ASD(A)	Assistant Secretary of Defense for Acquisition
ATD	Association for Talent Development
AWF	acquisition workforce
AWQI	Acquisition Workforce Qualification Initiative
BCE	Business-Cost Estimating
BFM	Business-Financial Management
CAP	Critical Acquisition Position
CHCO	Chief Human Capital Officer
CLM	continuous learning module
CON	Contracting
COTS	commercial off-the-shelf
CPCM	Certified Professional Contract Manager
DACM	Director for Acquisition Career Management
DAU	Defense Acquisition University
DAWDF	Defense Acquisition Workforce Development Fund
DAWIA	Defense Acquisition Workforce Improvement Act
DCAT	Defense Competency Assessment Tool

DCPAS	Defense Civilian Personnel Advisory Service
DCPDS	Defense Civilian Personnel Data System
DMSMS	Diminishing Manufacturing Sources and Material Shortages
DoD	U.S. Department of Defense
DoDI	Department of Defense Instruction
ECQ	executive core qualification
EDA	Executive Development Associates
ENG	Engineering
EVM	earned value management
FE	Facilities Engineering
FFRDC	Federally Funded Research and Development Center
FIPT	functional integrated product team
FY	fiscal year
GAO	Government Accountability Office
GSBPP	Graduate School of Business and Public Policy
HCI	Human Capital Initiatives
ICPM	Industrial Contract and Property Management
IT	information technology
ITEP	Information Technology Exchange Program
KLP	key leadership position
LCL	life cycle logistics
MBA	Master of Business Administration
MSCM	Master of Science in Contract Management
NCMA	National Contract Management Association
NDAA	National Defense Authorization Act
NDRI	National Defense Research Institute
NDS	National Defense Strategy
NDU	National Defense University

NGA	National Geospatial-Intelligence Agency
NPS	Naval Postgraduate School
OJT	on-the-job training
OPM	Office of Personnel Management
PM	Program Management
PMI	Project Management Institute
PMP	Project Management Professional
PP&C	Production Planning and Control
PQM	Production, Quality, and Manufacturing
PSC	Professional Services Council
PUR	Purchasing
RMF	Risk Management Framework
ROI	return on investment
SAF/AQXE	Air Force Industrial Liaison Office
SES	Senior Executive Service
SHCP	Strategic Human Capital Planning
SHRM	Society for Human Resource Management
SME	subject-matter expert
STM	Science and Technology Management
T&D	training and development
T&E	test and evaluation
USAASC	U.S. Army Acquisition Support Center
USD (AT&L)	Under Secretary of Defense for Acquisition, Technology, and Logistics
USD (A&S)	Under Secretary of Defense for Acquisition and Sustainment
USD (P&R)	Under Secretary of Defense for Personnel and Readiness
USD (R&E)	Under Secretary of Defense for Research and Engineering
WMG	Workforce Management Group

CHAPTER ONE

Introduction and Approach

Study Background

The Acquisition Workforce

The U.S. Department of Defense's (DoD) acquisition workforce (AWF) is "generally defined as uniformed and civilian government personnel, who are responsible for identifying, developing, buying, and managing goods and services to support the military."[1] The AWF is governed by statute,[2] and DoD is charged with developing and managing a highly skilled AWF that is managed "in a manner that complements and reinforces the management of the defense acquisition system."[3] As of June 2018, slightly more than 169,000 personnel—9 percent of them military—were considered to be part of the AWF, and they were distributed among 14 career fields, as shown in Table 1.1. The largest career field is Engineering, with 43,580 personnel, and the smallest is Property Management, with 391. Military personnel are most prominently represented in the Program Management career field, where 4,903 out of 17,727 (28 percent) are in the military.

Various statutes and DoD instructions govern the training, development, and management of the AWF. One of the most important is the Defense Acquisition Workforce Improvement Act (DAWIA), which was signed into law in 1990. DAWIA required that DoD "establish education and training standards, requirements, and

---

[1] Schwartz et al., 2016. They actually reference a RAND report for this definition: Gates et al., 2008, p. 2. Schwartz suggests a more detailed definition as follows: "Generally, the acquisition workforce consists of uniformed and civilian personnel who are either

- in positions designated as part of the acquisition workforce under the Defense Acquisition Workforce Improvement Act (10 U.S.C. §1721);
- in positions designated as part of the acquisition workforce by the heads of the relevant military component, pursuant to DOD Instruction 5000.66; or
- temporary members of the acquisition workforce or personnel who contribute significantly to the process, as defined in the Defense Acquisition Workforce Development Fund (10 U.S.C. §1705)."

[2] 10 U.S.C. Chapter 87, *Defense Acquisition Workforce.*

[3] 10 U.S.C. paragraph 1701a, *Management for Acquisition Workforce Excellence.*

Table 1.1
Military and Civilian Acquisition Workforce Demographics

Career Field	Military	Civilian	Total
Auditing	0	4,209	4,209
Business-Cost Estimating	74	1,360	1,434
Business-Financial Management	158	6,554	6,712
Contracting	4,444	26,304	30,748
Engineering	1,512	42,068	43,580
Facilities Engineering	4	11,133	11,137
Information Technology	216	7,384	7,600
Life Cycle Logistics	1,251	19,257	20,508
Production, Quality, and Manufacturing	741	10,149	10,890
Program Management	4,903	12,824	17,727
Property Management	0	391	391
Purchasing	0	1,321	1,321
Science and Technology Management	484	3,515	3,977
Test and Evaluation	1,880	6,927	8,807
Total	**15,667**	**153,396**	**169,063**

SOURCE: HCI, "Workforce Metrics for FY18(Q3)," 2018b.
NOTE: The Science and Technology Management distribution is approximate; the website did not break this career field down by military and civilian personnel.

courses" for both the civilian and military acquisition workforce, and it has been modified several times over the years.[4]

DAWIA established three levels of certification for acquisition positions, which are based on "their complexity, authority, and impact on defense acquisition programs, and not solely on a position's grade or rank":[5]

- Basic (Level I). Basic certification standards are reflective of fundamental competencies for the position. In addition to participating in education and training courses, individuals are expected to develop their required competencies through relevant on-the-job experience, including rotational assignments.

---

[4]   AcqNotes, "PBE Process: Defense Acquisition Workforce," last updated May 30, 2018.

[5]   U.S. Department of Defense (DoD), *Defense Acquisition Workforce Education, Training, Experience, and Career Development Program*, DoDI 5000.66, July 27, 2017a.

- Intermediate (Level II). Competencies at the intermediate level emphasize functional specialization. Individuals at this level are expected to have and apply journeyman-level acquisition-related skills. Broadening experiences provide the competencies and skills necessary to assume positions of greater responsibility. This may involve multifunctional experience and development.
- Advanced (Level III). This level is typically assigned to positions located in DoD components' organizations with a primary acquisition mission and where the duties require a high level of acquisition knowledge and skills.

DAWIA certification requirements for each career field are determined through a process managed by the career-field functional leader working through its functional integrated product team (FIPT), and the certifications themselves are granted by individual military services and the Fourth Estate. DAU publishes the DAWIA certification requirements and provides "a full range of basic, intermediate, and advanced certification training, assignment-specific training, applied research, and continuous learning opportunities" to AWF personnel so they can satisfy the certification requirements for the acquisition positions they fill or hope to fill.[6]

Some acquisition positions have requirements beyond DAWIA certification levels. Critical Acquisition Positions (CAPs) are positions that have "significant supervisory, managerial, or lead acquisition responsibilities." In addition to being Level III certified, personnel in CAPs "require tenure in order to ensure stability and provide accountability for the acquisition program, effort, or function, and must be filled by military officers at the O-5 grade or higher or civilians at the GS-14 grade or higher (and equivalent)." Other positions, designated Key Leadership Positions (KLPs), are even more demanding, with a "significant level of responsibility and authority that are key to the success of a program or effort. These positions warrant special management attention and oversight for qualification and tenure requirements."[7]

Congressional Interest in AWF Business Acumen, Knowledge of Industry Operations, and Knowledge of Industry Motivation

In addition to DAWIA, Congress has passed other legislation to improve the management of the AWF. The Defense Acquisition Workforce Development Fund (DAWDF) was created by the Fiscal Year (FY) 2008 National Defense Authorization Act (NDAA) as a "human capital tool for acquisition leaders across DoD to strategically bolster AWF recruiting, training and development (T&D), and retention efforts."[8] In FY 2010,

---

[6] See DoDI 5000.66 and Defense Acquisition University (DAU), *Defense Acquisition University 2018 Catalog*, undated(f), p. 14.

[7] Descriptions of CAPs and KLPs are in DoD, DoDI 5000.66, 2017a.

[8] See HCI (Office of the Under Secretary of Defense for Acquisition & Sustainment, Human Capital Initiatives), *Department of Defense Acquisition Workforce Development Fund: 2017 Year-in-Review Report*, March 7, 2018c.

Congress required DoD to develop strategic workforce plans for the AWF every two years, though this requirement was eliminated in the FY 2017 NDAA.[9]

Congress's continuing interest in developing a highly skilled AWF and in the T&D approaches used to do so is reflected in Section 843(c) of the FY 18 NDAA.[10] The actual legislative language is provided in Box 1.1. The act directed the Under Secretary of Defense for Acquisition and Sustainment (USD[A&S]) to assess AWF training, focusing on opportunities provided by industry and other external training sources in particular and gaps in business acumen, knowledge of industry operations, and knowledge of industry motivation.

The act mandated four specific areas to include in the assessment and the resultant report:

1.   current sources of T&D opportunities related to business acumen, knowledge of industry operations, and knowledge of industry motivation

---

Box 1.1
FY 2018 NDAA Section 843(c) Text

The Under Secretary of Defense for Acquisition and Sustainment shall conduct an assessment of the following:

(A) The effectiveness of industry certifications, other industry training programs, including fellowships, and training and education programs at educational institutions outside of the Defense Acquisition University available to defense acquisition workforce personnel.

(B) Gaps in knowledge of industry operations, industry motivation, and business acumen in the acquisition workforce.

(2) REPORT.—Not later than December 31, 2018, the Under Secretary shall submit to the Committees on Armed Services of the Senate and the House of Representatives a report containing the results of the assessment conducted under this subsection.

(3) ELEMENTS.—The assessment and report under paragraphs (1) and (2) shall address the following:

(A) Current sources of training and career development opportunities, industry rotations, and other career development opportunities related to knowledge of industry operations, industry motivation, and business acumen for each acquisition position, as designated under section 1721 of title 10, United States Code.

(B) Gaps in training, industry rotations, and other career development opportunities related to knowledge of industry operations, industry motivation, and business acumen for each such acquisition position.

(C) Plans to address those gaps for each such acquisition position.

(D) Consideration of the role industry-taught classes and classes taught at educational institutions outside of the Defense Acquisition University could play in addressing gaps.

SOURCE: Public Law 115-91, Fiscal Year 2018 National Defense Authorization Act (NDAA), December 12, 2017 (131 STAT.1480).

---

[9]   See DoD, *Acquisition Workforce Strategic Plan: FY 2016–FY 2021*, 2015, for the latest version of the Acquisition Workforce Strategic Plan.

[10]   An excellent summary of efforts through 2016 to improve the AWF is Porter et al., *Independent Study of Implementation of Defense Acquisition Workforce Improvement Efforts*, CNA, December 2016.

2. knowledge gaps related to business acumen, industry operations, and industry motivation for each acquisition position
3. plans to address those gaps for each acquisition position
4. consideration of the role that organizations outside of the Defense Acquisition University (DAU) could play in addressing gaps.[11]

USD(A&S) asked RAND's National Defense Research Institute (NDRI), a federally funded research and development center, to conduct the assessment.

Research Approach

In order to meet the congressionally mandated timeline, RAND researchers had six months to conduct and document their assessment. Work commenced in April 2018, and the draft report was completed in October 2018. In coordination with the project sponsor, the RAND team developed five tasks that would enable completion of the assessment directed by Congress within this compressed time frame:

- Task 1: Identify "in-use" T&D opportunities offered by DAU and industry providers. This included an examination of how AWF learning requirements are developed and communicated, how the need for the Section 843–related knowledge is documented, and which DoD and non-DoD sources are used to confer this knowledge to avoid or close gaps.
- Task 2: Identify known AWF knowledge gaps related to business acumen, industry operations, and industry motivation.
- Task 3: Conduct an environment scan of the landscape of external T&D options to identify additional external T&D opportunities not currently in use by DoD.
- Task 4: Consider the effectiveness of external T&D opportunities offered by industry and external educational providers. This included investigating approaches to evaluating T&D more generally that might help DoD decide the appropriate T&D mix to use.
- Task 5: Develop recommendations.

RAND researchers used multiple approaches to accomplish these tasks, including interviews with subject-matter experts (SMEs) inside and outside DoD; targeted

---

[11] Section 843 states that this training is to be considered for acquisition positions, as designated in Section 1721 of Title 10 U.S.C. It lists the following 12 types of acquisition *position*: program management; systems planning, research, development, engineering, and testing; procurement, including contracting; industrial property management; logistics; quality control and assurance; manufacturing and production; business, cost estimating, financial management, and auditing; education, training, and career development; construction; joint development and production with other government agencies and foreign countries; and intellectual property.

For reasons enumerated in this chapter, we will focus on acquisition workforce *career fields*.

discussions; a literature review of relevant studies and guidance; reviews of documents and websites categorizing DAU training, other DoD training, and "external" (non-DoD) offerings; a review of approaches used by DoD and non-DoD organizations to determine training effectiveness; and a review of AWF knowledge requirements (such as career-field competency models).[12]

Interviews

We conducted 44 interviews with DoD senior leaders and SMEs and with external stakeholders. A breakdown of these interviews follows:

- 26 interviews with DoD senior leaders and SMEs, including
  - directors (or acting directors) from all five of DAU's centers
  - directors for acquisition career management (DACMs) for the three military departments and DoD's Fourth Estate[13] (four interviews)
  - functional leaders (or their designees for interview purposes) for all of the AWF career fields (12 interviews)
  - DoD HCI leadership (one interview)
  - additional DoD personnel from DAU and the military departments (four interviews)
- seven interviews with representatives of professional associations
- five interviews with representatives of three private-sector organizations with industry rotation programs and/or extensive in-house corporate universities
- four interviews with representatives of three universities that were providing customized courses for AWF personnel at the time of this study
- two interviews with members of the Section 809 panel, a congressionally mandated panel tasked to identify ways to improve the defense acquisition process.[14]

We used a semi-structured approach for these interviews, which means that our interview protocol set forth opening questions and clear instructions but provided us with the discretion to delve into potentially fruitful lines of inquiry as they surfaced. The semi-structured nature of the interviews also meant that we posed different questions to different interviewees, and some remarks were elicited, while others were

---

[12] This study completed all necessary RAND administrative processes related to human subjects protection and received the determination that it met DoD's "not human subjects research" definition.

[13] *The Fourth Estate* is a term used to refer to the Office of the Secretary of Defense, defense agencies, and defense field activities outside of the three military departments and the unified combatant commands (Kathleen J. McInnis, *Defense Primer: The Department of Defense*, Washington, D.C.: Congressional Research Service, 2016).

[14] The Section 809 panel was initially called for in FY 2016 NDAA Section 809 and later amended by FY 2017 NDAA Section 863(d) and FY 2018 NDAA Sections 803(c) and 883. For more information, see Section 809 Panel, homepage, 2019.

shared spontaneously. Table 1.2 provides a list of major topics covered in the interviews, broken down by DoD and external interviewees. Most notably, in all of the interviews, we asked about perceived gaps in knowledge of business acumen, industry operations, and industry motivation for AWF personnel and extensively discussed current and potential uses of external T&D options to provide those types of knowledge. For the DoD interviews, we probed more deeply into career field–level issues and DoD processes, such as those related to competency models, gap assessments, T&D tracking, and evaluation, with questions that varied depending on the interviewee's position (e.g., we covered the same topics differently for functional leaders and DACMs).

Interviews were audio-recorded, professionally transcribed, and subsequently analyzed using a computer-assisted qualitative data analysis procedure referred to as "coding." Codes are applied in this data-reduction process to retrieve and organize qualitative data by topic and other characteristics. We employed a "structural" coding approach for this study; codes were based on our study goals and interview questions and were intended to help us identify themes. The use of a purposive, nonrandom sample (i.e., interviewees selected by virtue of their position and, in the case of external stakeholders, their availability during the data collection time frame) and the semi-structured nature of our interviews suggest that it would not be appropriate to either solely base or report findings via interview counts. Instead, we considered strength of interview evidence when identifying themes, meaning that we took into account

Table 1.2
Breakdown of Interview Topics by Interviewee Type

Topics	DoD Interviewees	External Interviewees
Section 843 knowledge type definitions (business acumen, industry operations, and industry motivation)	✓	
Perceived needs for Section 843 knowledge types	✓	✓
AWF career-field knowledge and competency requirements	✓	
T&D activities that confer the types of knowledge cited in Section 843	✓	✓
Perceived gaps related to Section 843 knowledge types	✓	✓
Role of external T&D in closing gaps	✓	✓
Facilitators of and barriers to DoD use of external T&D options	✓	
Processes related to communicating and tracking T&D activities	✓	
Processes related to evaluation of T&D	✓	✓
Recommendations to close or avoid gaps in knowledge related to business acumen, industry operations, and industry motivation	✓	✓

not only how frequently a topic was coded but also the richness of the discussion and the level of agreement across interviewees regarding a specific topic or theme. A more extensive discussion of our interview data analysis strategy is provided in Appendix A.

Targeted Discussions

We also had a series of discussions via telephone and email with DoD SMEs to learn more about DoD T&D efforts in use at the time of this study. These differed from the aforementioned interviews in that, rather than repeating similar questions across sessions, we asked a limited, unique set of questions relevant to the SME's background. The questions were intended to collect factual information only about specific aspects of DoD T&D. For example, we had targeted discussions with eight SMEs about

- Acquisition Workforce Qualification Initiative (AWQI) e-workbooks
- DAU's course-equivalency program
- industry attendance at DAU courses
- DoD efforts to collect standardized information about participants in industry rotation programs
- new efforts to develop technical competency assessments using the Defense Competency Assessment Tool (DCAT).

These conversations were not recorded; instead, one or more members of the RAND study team took notes.

Literature, Document, and Website Reviews

We reviewed DoD instructions (DoDIs) and documents that govern the development of competency models for AWF career fields and how these models are to be used by DAU for the development of performance objectives and learning assets.

To learn more about DoD approaches to providing learning opportunities, we reviewed the DAU catalog, DAU course descriptions, and directives addressing DAU's course-equivalency program. Based on information from interviews and DACM websites about other DoD sources of training, we reviewed documentation related to degree-granting opportunities, executive training options, and individual courses provided by other DoD institutions, such as the Naval Postgraduate School (NPS) and the Air Force Institute of Technology (AFIT).

For external (non-DoD) T&D opportunities, we examined online and published information on degree-granting programs, executive training programs, and training-with-industry programs available to AWF members.

We searched for published studies about the AWF, such as those completed by the U.S. Government Accountability Office (GAO), DoD, and Federally Funded Research and Development Centers (FFRDCs), focusing on those that addressed AWF competencies and training opportunities. We also asked those we interviewed for references or access to studies that assessed knowledge gaps in the AWF. This included both publicly available documents that our search queries might have missed and those they were aware of that were not publicly available.

To learn more about potential external resources for training that may confer business acumen and knowledge of industry, we focused on organizations mentioned in interviews with SMEs in DoD and elsewhere as offering potentially relevant examples. For each organization, we looked for information on the organization's practices around T&D. We also conducted a literature search around two specific topics of interest: evidence on best practices for the design and delivery of training and evidence on best practices for the evaluation of training. To identify literature, a range of different resources was used, including the Google Scholar and Business Source Direct search engines, as well as scans of content on the Harvard Business Review, Association for Talent Development (ATD), and Society for Human Resource Management (SHRM) websites.

AWF Knowledge Requirements Review

We reviewed the most up-to-date career-field competency models we could obtain from functional leaders, as well as lists of competencies provided in AWQI e-workbooks. As we will discuss further in Chapters Two and Three, we analyzed the models and workbooks to determine the extent to which they included competency elements related to the three types of knowledge specified in Section 843. We also reviewed the DAU course requirements for career-field DAWIA certifications to see how needs for these types of knowledge were reflected in required courses.

Table 1.3 shows how the various methods applied to different project tasks. Overall, each task was informed by multiple data sources.

While accomplishing these tasks, we maintained a career-field orientation for two reasons. First, the governing instruction for the development of competency models directs that the models will be developed by functional community.[15] Second, we hypothesized that the need for business acumen, knowledge of industry operations, and knowledge of industry motivation might vary across career fields and that these variations would affect the demand for training in the areas of interest to Congress. We also thought that these needs might vary by certification level, time in service, and individual status (military or civilian), and we explored these sources of variation, albeit less systematically than our career field–level analyses.

Variation of AWF characteristics is exhibited in Figure 1.1. The figure displays AWF career fields from largest (Engineering) at the top to smallest (Property Management) at the bottom and shows the distribution of Level I, Level, II, and Level III positions. CAP and KLP positions require Level III certification, and their distribution

---

[15] DoD, DoDI 5000.66, 2017a. Functional leaders are SMEs for their respective functional and competency areas and provide senior oversight to one or more acquisition career fields.

Table 1.3
Applicability of Research Methods to Project Tasks

Method	Task 1: In-Use Options	Task 2: Knowledge Gaps	Task 3: Environment Scan of External Options	Task 4: Gauging Effectiveness
Interviews with DoD stakeholders and SMEs	✓	✓	✓	✓
Interviews with external stakeholders	✓	✓	✓	✓
Targeted discussions with DoD SMEs	✓			
Review of competency models and other sources of knowledge requirements		✓		
Review of DAU offerings	✓			
Review of other DoD-based T&D options (e.g., service schools)	✓			
Review of external T&D options (e.g., those offered by industry, external universities, and commercial vendors)	✓		✓	
Review of relevant reports and studies (e.g., GAO, CNA, RAND)		✓	✓	✓
Review of in-use effectiveness approaches				✓

is highlighted in the chart as well. Overall, there are 16,451 CAP positions and 1,076 KLP positions.

Key Definitions Used in Our Analysis

While Congress expressed an interest in determining gaps in business acumen, knowledge of industry operations, and knowledge of industry motivation present within the AWF, the NDAA language does not define any of these terms. As a result, we searched DoD and Office of Personnel Management (OPM) documents for possible definitions, and we asked those we interviewed whether they were aware of any official definitions for these terms and, if not, how they would define them.

We were only able to find an official government definition for knowledge related to *business acumen*. OPM has identified five executive core qualifications (ECQs) for the government's Senior Executive Service (SES) positions,[16] and business acumen is the fourth. It is defined as "the ability to manage human, financial, and information resources strategically," with those three areas defined as follows:[17]

---

[16] OPM, "Senior Executive Service: Executive Core Qualifications, undated.

[17] OPM, undated.

Figure 1.1
Acquisition Skill–Level Workforce Demographics

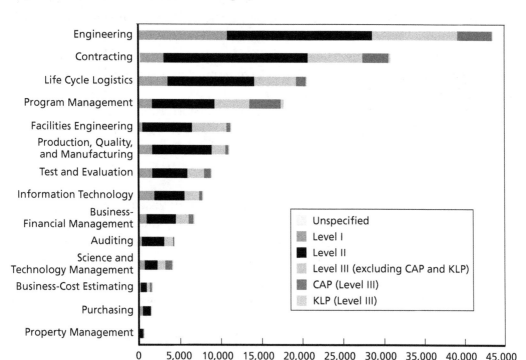

SOURCE: HCI, 2018b.

- "Financial Management: Understands the organization's financial processes. Prepares, justifies, and administers the program budget. Oversees procurement and contracting to achieve desired results. Monitors expenditures and uses cost-benefit thinking to set priorities.
- Human Capital Management: Builds and manages workforce based on organizational goals, budget considerations, and staffing needs. Ensures that employees are appropriately recruited, selected, appraised, and rewarded; takes action to address performance problems. Manages a multi-sector workforce and a variety of work situations.
- Technology Management: Keeps up-to-date on technological developments. Makes effective use of technology to achieve results. Ensures access to and security of technology systems."[18]

---

[18] One of our reviewers noted a resource that did not emerge from our interviews or search for formal definitions: DoD Instruction 1430.16, *Growing Civilian Leaders*, 2009, which includes business acumen as a competency for DoD leaders. It mentions the three components in the OPM as well as a fourth area, computer literacy.

Insights from our interviews about all three types of Section 843–related knowledge and the limited information we found related to business acumen—namely, the ECQ definition of that term—led us to develop the definitions shown in Figure 1.2, which were approved by our research sponsor for this project.

Note that in the case of business acumen, we expanded on the OPM definition by including an understanding of industry behavior and trends that enables one to shape smart business decisions for the government.

These definitions were important for our categorization of AWF knowledge requirements, analysis of the content of various DAU courses, and determination of the existence of gaps related to these three types of knowledge.

Research Limitations

A few limitations of this research, some related to the compressed schedule required to meet the congressional deadline, are important to note at the outset of our report. First, the limited scope of this study meant that we were unable to collect data via interviews or surveys with AWF workforce members. The administrative approval processes this would have entailed and the time and effort it would have required to recruit, collect, and analyze proficiency ratings from or about defense acquisition personnel rendered such efforts infeasible.

In addition, the lack of official definitions for *business acumen*, *industry operations*, and *industry motivation* meant that we had to develop our own before we could assess knowledge requirements and gaps related to those three types of knowledge, and differing perceptions among those we interviewed of the meaning of these terms could have affected how they categorized gaps in knowledge. Our analyses of the need for knowledge in these three areas, as expressed in competency models and DAU courses

Figure 1.2
Definitions of Business Acumen, Industry Operations, and Industry Motivation

Business acumen	Industry operations	Industry motivation
In addition to the ability to manage human, financial, and information resources strategically (OPM definition), business acumen is an understanding of industry behavior and trends that enables one to shape smart business decisions for the government.	This includes plans and procedures used within an industry to provide a product or service. The need for knowledge of specific practices may vary, depending on an employee's contribution to the acquisition mission. Some industry operations may be business oriented, while others may be at the confluence of business and technical knowledge—i.e., "techno-business" (e.g., milestone reviews).	This includes the range of considerations and motivations that factor into the decisionmaking of organizations in industry, including profit and revenue, market share, management and employee incentives, shareholder considerations, perspectives on risk, and the need to maintain position in a competitive environment. The relative weights of these factors may vary by industry and over time.

required for DAWIA certification, is dependent on the definitions we developed, and we recognize that different definitions would lead to different assessments.

The time available for the work meant that we could not conduct as many expert interviews as we otherwise would have. For example, we learned about but did not have the opportunity to pursue interviews with service-level career-field managers and service-specific schools. Also, we experienced difficulties recruiting industry representatives, particularly those from commercial firms, and the study time frame prevented us from engaging in multiple rounds of identifying and recruiting industry-based interview participants. Fortunately, we were able to interview representatives from both traditional defense contractors and technology firms, including ones with in-house corporate universities.

Some of the data we sought were also unavailable. There was an unexpected dearth of preexisting studies on knowledge gaps within the AWF, for example. In addition, we were unable to obtain T&D-related statistics such as historical information on the number and type of participants in industry rotations and the number of AWF members who have attained business-related graduate degrees.[19] Finally, as we will discuss further in the report, competency models for the DCAT technical competency update are still in development, so we could not examine them for their coverage of Section 843–related knowledge or compare them with existing career field–level competency models.

While these limitations were unfortunate, they also helped highlight areas related to the management of competency models and training options that warrant improvement. We worked to compensate for these shortcomings by triangulating findings from multiple data sources to the greatest extent possible.

Organization of the Report

Chapter Two outlines the formal process that DoD uses to define competencies and the need for knowledge of business acumen, industry operations, and industry motivation and reports interviewee assessments of the need for these types of knowledge. Chapter Three describes how DoD identifies knowledge gaps within the AWF, the approach we took to determine gaps in the types of knowledge cited in Section 843, and the results of our assessment. Chapter Four shows the various internal (DoD) and external (non-DoD) resources that are used to provide T&D for the AWF, discusses interview insights into the potential for greater use of external T&D resources to provide the knowledge focused on in Section 843, and outlines a portfolio approach to addressing T&D gaps. Chapter Five presents the results of our investigation of ways to

---

[19]   An exception in the case of industry rotations was the Air Force. The Education with Industry program manager had detailed, verified summaries of participation for the past five years.

assess the effectiveness of external T&D in closing gaps related to knowledge of business acumen, industry operations, or industry motivation. Chapter Six summarizes our conclusions and recommendations. The report also features two technical appendixes: Appendix A details our interview methodology, and Appendix B describes our approach to analyzing competency models and DAU courses.

CHAPTER TWO

The Need for Business Acumen, Knowledge of Industry Operations, and Knowledge of Industry Motivation

In this chapter, we first delineate the formal process prescribed in DoD guidance for developing AWF competency models and translating those models into T&D, such as courses of instruction at DAU. Next, we discuss how this process has been implemented in practice and variation across the AWF career-field–level competency models. We then analyze the extent to which the business-related knowledge needs highlighted in Section 843 are captured in the AWF career-field–level competency models, how they appear in the DAU courses required for DAWIA certification, and how these needs vary by career field. Finally, we describe perceptions of those we interviewed of the needs for business acumen, knowledge of industry operations, and knowledge of industry motivation by career field and whether needs might vary by career stage or certification level.

DoD Competency Models

Regulatory Background

DoD guidance calls for the use of a competency management framework for its strategic human capital planning, in which a *competency* is defined as "an observable, measurable pattern of knowledge, abilities, skills, and other characteristics that individuals need to perform work roles or occupational functions successfully."[1] The aim of this framework is to align mission outcomes with expectations for employee behaviors, "providing a meaningful and consistent structure within which to define and assess workforce competency needs and gaps, and providing employees and supervisors with observable, transparent, and measurable indicators associated with successful job performance."[2]

The framework seeks to achieve this aim by

---

[1]   DoD, DoDI 1400.25, 2016a.

[2]   DoD, DoDI 1400.25, 2016a.

- providing a common language and structure to assess competency gaps and proficiency levels
- establishing an inventory of competencies by occupation
- establishing a common taxonomy for DoD-wide competency management in a way that leads to standardization but also allows for flexibility if needed.

The framework has five tiers of competencies:[3]

- Tier 1: Core competencies. These are DoD-wide competencies, such as DoD leadership competencies.
- Tier 2: Primary occupational competencies. These are competencies for an occupational series or function, such as acquisition career fields.
- Tier 3: Sub-occupational specialty competencies. These apply to specialties within an occupation, such as a subset of civil engineers.
- Tier 4: DoD component–unique competencies. These are unique to a specific component, such as a military service.
- Tier 5: Position-specific competencies. These are competencies required for a certain position that are not captured in the other tiers of competencies.

The career-field competency models we review in this chapter are examples of Tier 2 competency models. The framework also includes a competency taxonomy consisting of five levels of proficiency that are tied to employee performance and assessments: (1) awareness, (2) basic, (3) intermediate, (4) advanced, and (5) expert.[4]

Those tasked with AWF management have adopted this framework and documented the process that links competency models to proficiency standards and learning objectives for AWF personnel in DoDI 5000.66, *Defense Acquisition Workforce Education, Training, Experience, and Career Development Program*.[5] Specifically, this instruction

- describes the roles and responsibilities of key DoD officials and entities
- discusses how the competency management framework is operationalized in the context of the AWF[6]
- outlines the different types of AWF positions and indicates additional requirements that apply to individuals in certain roles, such as CAPs and KLPs
- provides an overview of the career-field certification process.

---

[3]   DoD, DoDI 1400.25, 2016a.

[4]   DoD, DoDI 1400.25, 2016a.

[5]   DoD, DoDI 5000.66, 2017a.

[6]   In addition to career-field models (our focus), competency models also exist for career paths that can fall within or cut across career fields. Career path models are to be written at the Tier 3 level. See DoDI 5000.66, 2017a.

Competency-Model Development

DoDI 5000.66 calls for the Under Secretary of Defense for Acquisition, Technology, and Logistics (USD [AT&L]) to designate functional leaders to provide "senior oversight to one or more acquisition career fields or career paths."[7] DoD currently has 11 functional leaders who "serve as subject matter experts for their respective functional and competency areas."[8] Functional leaders organize FIPTs to support them.

> FIPTs are usually led by a designated person at the GS-15 level. Each Service has two representatives on the FIPT: a technical person (e.g., a contracting expert) and a manpower person representing the DACM. This person usually sits on the FIPTs for two or three acquisition career fields. (DoD 10)[9]

Through their FIPTs, functional leaders develop competency models, and coordinate with the DAU Capabilities Integration Centers to "define the knowledge, skills, and abilities (sub-competencies) that comprise the competency models, including proficiency standards, learning objectives, and other talent management applications, as appropriate."[10] According to the FY 2016–2021 AWF Strategic Plan, the goal is to conduct career-field competency assessments every five years.[11] In addition, DoDI 5000.66 states that "Functional Leaders will annually validate, update, and approve the models, as required."[12]

---

[7]   DoD, DoDI 5000.66, 2017a. As of February 1, 2018, the former USD(AT&L) has been restructured into two offices, per provisions in the FY 2017 NDAA: the Under Secretary of Defense for Acquisition and Sustainment (USD[A&S]) and the Under Secretary of Defense for Research and Engineering (USD[R&E]). The reorganization process has not yet been completed; however, a preliminary document outlining how the division of responsibilities was to occur listed "Acquisition Workforce Policy/Training" among the responsibilities of the USD(A&S). See U.S. Department of Defense, *Report to Congress: Restructuring the Department of Defense Acquisition, Technology and Logistics Organization and Chief Management Officer Organization*, Washington, D.C., August 1, 2017b.

[8]   DoD, DoDI 5000.66, 2017a.

[9]   Because we assured interview participants that we would not attribute findings to a specific individual, after each quotation, we indicate whether the participant is representing a DoD organization or position (e.g., DAU faculty, functional leader) or industry (e.g., external educational institution, private-sector company, professional association). As we prepared the report, following Guest et al. (2011), each participant was assigned a unique identifier that included both a number and his or her organization's sector to ensure that we were not serially quoting any single individual. For a discussion of this practice, see G. Guest, K. M. MacQueen, and E. E. Namey, *Applied Thematic Analysis*, Thousand Oaks, Calif.: SAGE Publications, 2011, pp. 267–268. For some interviews, we redacted potentially identifying details to protect participant confidentiality.

[10]   DoD, DoDI 5000.66, 2017a.

[11]   DoD, *Acquisition Workforce Strategic Plan: FY2016–2021*, Washington, D.C., undated, p. 26.

[12]   In an October 15, 2018, telephone discussion with representatives from different functional areas, we were told that this annual review is a means to verify that current competencies are still satisfactory and that they are being addressed by DAU. It is not meant to be a formal competency assessment. In a December 18, 2018, e-mail, DAU's Director of Academic Programs said that DAU began requesting this annual validation in FY 2008. Typically, DAU sends a request for validation to functional leaders in April (along with a suggested template for a

Figure 2.1 illustrates the stages of the process. As indicated in the figure, on the behalf of functional leaders, FIPTs endeavor to develop competency models for their respective career fields. After the models are completed, functional leaders work with DAU to distill from the models necessary proficiency standards and learning objectives. DAU, in turn, is responsible for developing the "learning assets"—such as DAU classes that are part of the DAWIA certification process—that enable AWF members to develop the competencies required by their respective career-field models.[13]

The Acquisition Workforce Qualification Initiative

As part of the Better Buying Power 2.0 initiative, USD(AT&L) released a memo in April 2013 that directed all functional leaders to finalize competencies for each functional area by July 1, 2013. DAU was to translate those competencies into individual qualification plans that would allow each AWF member to tie his or her performance to the qualification. Tools to do this were to be developed by July 1, 2014,[14] and the effort was referred to as the Acquisition Workforce Qualification Initiative (AWQI).[15]

Figure 2.1
AWF Competency Model Framework

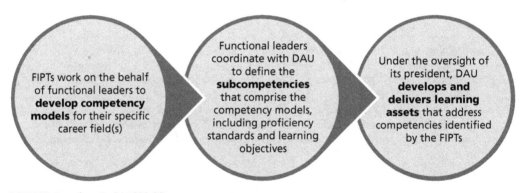

SOURCE: Based on DoDI 5000.66.

---

response) requesting that the certification be completed by May. This timeline satisfies DoDI 5000.66 requirements but also supports DAU catalog preparation and updates to DAU website information.

[13] The competency-model description in DoDI 5000.66 introduces terminology that is not in DoDI 1400.25. The AWF model has units of competence, which are composed of two or more competency topics. Competency topics comprise two or more competencies, and each competency can include two or more subcompetencies. DoDI 5000.66 says that the subcompetencies are the knowledge, skills, and abilities required for a position.

[14] USD(AT&L), *Implementation Directive for Better Buying Power 2.0—Achieving Greater Efficiency and Productivity in Defense Spending*, memorandum, April 24, 2013a.

[15] BBP 2.0 was announced in November 2012 (USD[AT&L], *Better Buying Power 2.0: Continuing the Pursuit for Greater Efficiency and Productivity in Defense Spending*, memorandum, November 13, 2012), but the implementation guidance was issued in April 2013 (Kendall, 2013a). Neither memo uses the term *AWQI*, but the pro-

A 2015 GAO report[16] commented on the difficulty of creating "a set of standards whose applicability would be common across all personnel, including those with the same position title, because employees perform different acquisition activities across or even within the DOD components,"[17] but it also noted that leaders of the AWQI were working on a spreadsheet-based tool that employees could use to help support career-development conversations with their supervisors.

AWQI e-workbooks are now available online[18] and are meant to help identify "on-the-job developmental opportunities and capture demonstrated acquisition experience." These workbooks have a standardized structure that links a unit of competence to a competency, a competency element, a product, and, finally, a task. For example, for one competency in the Production, Quality, and Manufacturing (PQM) career field, the e-workbook lists the following:

- Unit of competence: Defense Acquisition Management Process
- Competency: Knowledge of the DoD acquisition process, to include the DoD 5000 series and related policies
- Competency element: Knowledge of DoD processes for how systems evolve from mission needs through development and production to deployment and disposal
- Product: Develop the manufacturing strategy and quality management strategy in support of the acquisition strategy.
- Task One (out of five tasks): Obtain the acquisition strategy and determine production, quality, and manufacturing areas to be developed.

The AWQI e-workbooks exclude those competencies in the career-field competency models that the AWQI developers deemed to apply to the AWF and to non-acquisition personnel, such as leadership, communication, and the ability to work effectively with industry. The exclusion was intentional because it was difficult to define measurable tasks for these competencies or describe standards for them.[19]

---

cess is clearly what the AWQI people were doing. The leader of the AWQI effort confirmed in an October 16, 2018, phone call that work began in 2013.

[16]   GAO, *Defense Acquisition Workforce: Actions Needed to Guide Planning Efforts and Improve Workforce Capability*, GAO-16-80, Washington, D.C., December 2015.

[17]   GAO, 2015, pp. 19–20.

[18]   The AWQI workbooks are online at DAU, "Acquisition Workforce Qualification Initiative," undated(a).

[19]   Targeted discussions with DoD SME, June 21, 2018, and October 16, 2018. The SME provided us a separate workbook that listed these "non-acquisition-unique" competencies that are part of the career-field competency models but that were excluded from the AWQI e-workbooks.

Implementation of Career-Field Competency Models

Consistent with DoD guidance, competency models exist for all AWF career fields;[20] however, as we discuss in this section, inconsistencies remain—in particular, in how career fields have mapped those competencies to levels of proficiency and experience levels, if they have been mapped at all.

Competency Model Development Process in Practice

Interviewees mentioned numerous sources that inform the development of the career-field competency models by the functional leaders and FIPTs. These ranged from more-formal sources (e.g., legislative requirements and the National Defense Strategy) to studies conducted by internal or external entities (e.g., the Defense Science Board and the Systems Engineering Research Center) to informal conversations with people in the career field and lessons learned from experience.

> We talk to people who do the job. We talk to the people who have been here for a while and are at the senior levels and working this and said, you know, "What do you need in your everyday work?" (DoD 23)

> He's [the functional leader] not acting in a vacuum. He was taking inputs and concerns and recommendations from all of the service organizations. . . . They all know intuitively the workforce professional development requirements and expertise required of their own communities and they are going to feed that information right back into the functional IPT process. (DoD 8)

Feedback from DAU courses and reviews of related career fields' competency models can also factor into the development of the models.

> [E]very year we look through the student feedback on all the courses that we have at DAU and if they give us some good feedback, we'll feed that into the competency model. (DoD 4)

> [W]e keep track of what the Engineering career field does, because obviously they have a lot of similar competencies and every now and then, we'll see something that maybe we didn't think about that one and we'll talk about it. And if it seems appropriate, we will go ahead and add it. (DoD 20)

---

[20] All career fields, with the exception of Auditing, also are included in the AWQI e-workbooks. The Auditing career field is governed differently from the remainder of the career fields, with the AWF FIPT playing a lesser role in guiding the development of competencies and greater overlap with other segments of the DoD workforce that have their own knowledge requirements. For example, members of the Auditing AWF career field may also be part of DoD's Financial Management workforce (which overlaps but is distinct from the AWF career field by the same name), which has its own certification procedures, and also must adhere to GAO's "Yellow Book" of Generally Accepted Government Auditing Standards.

Use of Industry Materials to Inform Competency Model Development[21]

The use of industry certifications and industry-developed competency or knowledge requirements to inform the competency requirements tends to be done informally, with variation across the career fields. For example, during the interviews we heard remarks such as the following:

> We look to those kind of things like when we looked at the Project Management Certification, a lot of what we have in our curriculum reflects some of that. . . . So yeah, we do look to some of those—probably not formally. (DoD 2)

> [There] are industry bodies that have certifications like APICS and the Society for Logistics Management . . . but we haven't made those a required thing. . . . it isn't tied to the competency model. . . . it isn't an explicit thing that you have to have a . . . commercially recognized certification for promotion or retention. (DoD 15)

Some interviewees mentioned using materials developed by project management associations (e.g., Project Management Institute [PMI], International Project Management Association) or universities that offer project management certification to inform the development of the Program Management competency model. For example, one commented:

> We went through and we made sure that there wasn't anything left out, anything in the PMI-type certification that should be included in the Program Manager list of competencies. (DoD)[22]

Other DoD interviewees were less keen on incorporating industry-developed materials in the DoD competency models, citing reasons such as a hesitancy to appear to endorse a particular certification or set of materials to a lack of comparability between what industry does and the specialized DoD acquisition process. As one interviewee noted:

> [W]e tried to map that to the industry and what we found was there're a bunch of different ways the industry handles tests and evaluation and identifies those people and those skill sets in the different prime contractors. So, it's really hard on the industry side to equate the T&E competency model to a person or certification on the other side. (DoD 4)

---

[21]   This section addresses government sources of competencies. Various professional organizations relevant to the AWF also develop sets of competencies. For example, NCMA has developed its own Contract Management Body of Knowledge (CMBOK). PMI has developed a Project Management Body of Knowledge (PMBOK). APICS (formerly known as the American Production and Inventory Control Society) and the Society for Logistics Management also develop standards for certifications.

[22]   To protect the confidentiality of this interviewee, we have not included the numeric identifier.

Some interviewees acknowledged that they do not look to industry certifications at all, as evidenced by comments such as "I have not looked at industry certification competencies" (DoD 20). One interviewee expressed the view that DoD should not do so, stating: "[I]t's really difficult for me to say that we should be relying too much on industry input" (DoD 12). In contrast, other interviewees thought that more collaboration with industry may be in order:

> We have not engaged those professional certification societies a real whole lot in trying to drive what goes into our curriculum. Perhaps we should. (DoD 10)

For their part, several industry interviewees thought that there could and should be greater collaboration in advising and developing common competency and knowledge requirements. For example, one commented:

> I think, most of the competencies could be developed in tandem collaboratively and would be understood and remain understood the same way by both, which by the way makes it a whole lot easier in negotiating a contract. (Industry 17)

Another took this even further, suggesting that industry standards replace DoD ones:

> DoD standards are much narrower than NCMA [National Contract Management Association] standards. DoD standards focus only on the buyer's side, so they don't teach personnel about all of the things the contractor is doing as a seller. And you have to know the whole picture. NCMA standards cover both the buyer and the seller's perspective. I'm not sure why DoD doesn't just adopt NCMA's standards and insists on having its own. (Industry 5)

Use of Competency Assessments to Inform Competency Model Development

In-depth competency assessments occur about every five years, according to interviewees; are often conducted in tandem with outside organizations, such as FFRDCs, University-Affiliated Research Centers, or academic researchers; and involve reaching out to a majority of career-field members. Depending on the career field, such assessments are used to develop models or to inform revisions to them. The comments that follow convey how these processes are carried out:

> We go through every five years at a minimum to review those competencies and assess whether they are current and need to be updated. That is done through sending things out to the Acquisition Workforce and their supervisors having them do a survey to assess the credibility of each of the competencies that are currently listed. (DoD 19)

> I think the model [development work] really is like every four or five years and with 15 career fields, it takes us four or five years to get through them all. Because

to actually do a full competency assessment takes pretty darn close to a year. (DoD 26)

Less formal and less comprehensive competency assessments happen more regularly and can involve surveys, discussion forums, or unstructured conversations. Interviewees indicated that competency reviews are an ongoing, iterative process of assessing competencies and determining whether DAU courses align with them. According to some interviewees, the recertification of competency sets should happen annually.

> So we continuously look at . . . continuously . . . we do an assessment to see do we think we have the right competencies. We just completed an assessment and again, we're revising one of our DAU courses, because we think it's not as relevant as it needs to be. So we'll go through that revision. (DoD 14)

> I would also argue the OSD functionals do a review of their competency set annually. . . . And tell DAU whether or not they think it's accurate or whether or not there are gaps . . . and it can come from any level of feedback. Sometimes it comes from process change. Sometimes it comes from statute change. Sometimes it comes from just the subject-matter experts that are looking at that. They have to self-declare whether or not they've got the same set or whether or not DAU needs to adjust training. They have to do that annually. (DoD 7)

Competency assessments also can occur at the service-specific level or for a particular career field within a service. For example, the Army has conducted competency assessments of its AWF in accordance with its Army Acquisition Workforce Human Capital Strategic Plan.[23]

Current Career-Field Competency Models

Despite the formal guidance described above that seeks to standardize the career-field competency models, the models that we obtained from DoD, either directly from career-field functional leaders or from DAU, had inconsistencies in their structures, levels of detail, and how recently they appear to have been updated. This lack of standardization may reflect the fact that most of the competency models we received were developed before the publication of the 2016 versions of DoDI 1400.25 and DoDI 5000.66, which describe a more standardized approach. As noted in the discussion of AWQI, the AWQI e-workbooks do have a standard format (unit of competency, competency, competency element, product, task), but those workbooks exclude "non-acquisition-unique" competencies from the career-field competency models, some of which cover topics that have overlap with the types of business-related knowledge specified in Section 843. Thus, in order to analyze competency models for the career fields

---

[23] Sable, "AAW Human Capital Strategic Plan: Year One," *Army AL&T Magazine, Career Development, HCSP*, September 5, 2017.

for which we did not receive a model directly from the functional leaders (about half of them), we did so by separately considering the contents of the AWQI e-workbooks and the separately provided workbook of "non-acquisition-unique" competencies.[24]

There was variation across the career-field competency models we received. These models often appeared to be the sum of the AWQI (devoid of products and tasks) and the "non-acquisition-unique" competencies from the separate workbook, though our review of the models revealed exceptions: Most notably, two career fields (Program Management and Science and Technology Management) appear to have experienced significant competency model updates since the development of AWQI. With regard to taxonomy, rather than the standardized terms of unit of competency, competency, and competency element included in AWQI, the FIPT-produced models had different categorization schemes. One included unit of competency, topic, and competency; another had unit of competency, competency, and subcompetency; and another had topic and competency (but no lower level of disaggregation, such as a competency element or subcompetency). Moreover, the length and level of detail varied across the competency models; one model had more than 350 distinct competency elements, while most had fewer than 100, and one had just 25.

Only two of the career fields had competency models that included a hierarchy of proficiency levels, as is called for in DoD guidance. The Program Management model included competency element descriptions written at three levels—basic, intermediate, and advanced. Contracting is the only career field that had a model that included the five proficiency levels outlined in DoDI 1400.25—awareness, basic, intermediate, advanced, and expert.[25]

Other Government Sources of Competencies

The AWF career-field competency models are not the only federal sources of competencies. As described above, additional tiers of competencies include DoD-wide competencies and competencies pertaining to specialties within an occupation or a particular component. Interviewees noted the layers of competencies and how they can vary across the services and agencies. One stated that "position requirements may dictate additional knowledge elements and that may vary by DoD components" (DoD 22), while another commented:

---

[24] We also received from DAU a "Career Field Competency Baseline for RAND Study" workbook that included a competency model for the career fields for which we did not receive a model directly from the functional leaders; the models in this workbook were identical to those in AWQI except that they did not include the products and tasks that are layered onto the competency elements in AWQI; the only exception was Test and Evaluation, for which there were two competencies in this separate file that were not included in AWQI.

[25] While the Contracting competency model defines these five levels for most competencies, it does not define them for the ten "professional" competencies in the model (problem solving, customer service, oral communication, written communication, interpersonal skills, decisiveness, technical credibility, flexibility, resilience, and accountability).

Contracting, for example, there's an OSD set of competencies but if I go to the Air Force Contracting Functional, I'm sure they've got a different set that might go down deeper. And if I go to like a Contracting Squadron Commander, they might have local identified competencies or learning objectives they want folks to have to support their local mission. (DoD 7)

When it comes to the areas of knowledge highlighted by Section 843, two other sources of competencies—both of which cut across career fields and components—are important for those in AWF leadership positions.

The first source, which lies outside the DoD competency management framework, was mentioned in Chapter One in our discussion of the definition of business acumen: OPM has prescribed five executive core qualifications (ECQs) that are required for entry to the Senior Executive Service (SES).[26] As OPM notes, the ECQs are designed to assess executive experience and potential, not technical expertise. ECQ number four is business acumen, so members of the AWF who are in SES positions must have this qualification regardless of career field.

The second source is a 2013 memorandum from the USD(AT&L) that defined five essential factors for the selection of AWF personnel to KLPs: education, experience, cross-functional competencies, tenure (an agreement to remain in a position for a specified period), and currency in continuous-learning requirements. The cross-functional competency requirements include six executive competencies, and one of them is business acumen.[27]

Evidence of Section 843–Related Knowledge Needs, as Indicated by Career-Field Competency Models and DAWIA Certification Requirements

We used two complementary approaches to determine the extent to which existing sources of requirements indicate that acquisition workforce personnel in each career field need the three types of business-related knowledge specified in Section 843. We first reviewed career-field competency models to identify competencies that pertain to business acumen, knowledge of industry operations, and knowledge of industry motivation; next, we analyzed DAWIA requirements to see how many DAU courses that convey Section 843–related knowledge are required for certification; and, finally, we compared at a high level how well the two were aligned. The rest of this section briefly describes the methodology; Appendix B provides more details.

---

[26] OPM, undated.

[27] The executive leadership competencies are fundamental, leading change, leading people, results driven, business acumen, building coalitions, and enterprise-wide perspective. The first five are the same as OPM's ECQs.

Competency Models

We reviewed competency models for each career field to determine the extent to which the models indicate that personnel in the career field need the types of knowledge cited in Section 843.[28] In doing so, we sought to mitigate the impact of inconsistencies across the competency models in terms of their structure and level of detail by tabulating the number of categories of Section 843–related knowledge that are included in each career-field model rather than using a raw count of the number of competency model elements that involve this knowledge.[29]

We proceeded in an iterative manner, going through each competency model to flag competency elements with language related to business acumen, industry operations, or industry motivation, and tagging these competency elements with one or more keywords that appear in our definitions of these three terms or that came up in our interviews. We then went back through the models to ensure consistency in how the keyword tags were applied and to finalize a set of 16 categories that corresponded with the business-related knowledge types in Section 843. These 16 categories are listed in Table 2.1, along with an example of language from the competency models that we determined to be related to the keyword tag. Notably, "business acumen" appears explicitly in a few of the career-field competency models that we reviewed; however, the terms "industry operations" and "industry motivation" do not appear explicitly in any of them.

We tallied the number of categories of Section 843–related knowledge incorporated in each career field's competency model and calculated an average across the career-field models.[30] We deemed career fields with competency models that cover an above-average number of aspects of business acumen, knowledge of industry operations, and knowledge of industry motivation to have a "higher" relative need for these types of knowledge, as indicated by the competency models, while career fields with competency models that include a below-average number receive the "lower" relative need designation. Career fields with higher Section 843–related knowledge needs as indicated by the competency models include

---

[28] The competency models we reviewed were either those provided to us directly by career-field functional leaders or those embedded in the AWQI workbook. When we reviewed a competency model from the AWQI workbook, we did so in tandem with a review of competencies for the career field included in a workbook provided to us by an AWQI official that lists "non-acquisition-unique" competencies, which are competencies that are included in the career-field competency models but that were excluded from AWQI. The Business-Cost Estimating and Business-Financial Management career fields are consolidated into one "Business" competency model.

[29] For example, the Life Cycle Logistics career-field competency model that we reviewed includes 363 competency elements, compared with fewer than 100 competency elements in most other career-field models. To be sure, while our methodological approach weights the breadth of Section 843–related knowledge requirements over the sheer number of competency elements, longer, more detailed models may be more likely to include more of our categories of knowledge.

[30] The average number of our categories of Section 843–related knowledge included in the career-field competency models is 9.8, ranging from five to 14.

Table 2.1
Categories of Section 843–Related Knowledge Included in Competency Models

Category of Knowledge	Example from Competency Models
Technology/technical management	"Pursue self-development to advance technical and management skill sets and prepare for future advancement and changing technologies" (Test and Evaluation)
Human capital management	"Manage human capital, project funds, and laboratory/facility capital to insure project stays on budget and obligations and disbursements are timely" (Science and Technology Management)
Resource/risk management	"Risk Management" competency, which includes two competency elements (Industrial Contract and Property Management)
Negotiation	"Negotiate terms and conditions (including price) based on the pre-negotiation objective and give-and-take with the offeror to establish a fair and reasonable price" (Contracting)
Contractor costs and pricing	"Cost, Pricing, and Rates/Cost Management: Apply knowledge of the cost accounting basics defense companies use to manage direct and indirect costs and the use of rates to contribute to the preparation of appropriate acquisition strategies and solicitations, and to provide necessary technical oversight of contract execution" (Engineering)
Contractor performance	"Knowledge needed to assess supplier performance and determine supplier capability to ensure prime contractors control of subcontractors and vendors" (PQM)
Market research	"Conduct market research using relevant resources prior to solicitation to understand the industry environment and determine availability of sources of supply and/or services" (Purchasing)
Industry best practices	"Identify and implement commercial best practices for supply chain management" (Life Cycle Logistics)
Industry perspectives/ stakeholders	"Conduct pre-solicitation industry conferences and analyze responses to draft solicitation terms and conditions to promote full and open competition" (Contracting)
Incentives	"Recommend contractor/financial incentives to promote the contractor performance that would be in the best interest of the government" (Business)
Evaluating industry proposals	"Prepare and evaluate scopes of work and proposals for design build contracts for acquisition of facilities that result in projects that meet or exceed criteria, are under budget, and provide ahead of schedule" (Facilities Engineering)
Earned value management	"Earned value management (EVM)" competency, which includes two competency elements (Information Technology)
Small business	"Evaluate how to use small business resources during the planning process of the acquisition lifecycle phases" (Program Management)
Industry standards	"Support the use of commercial standards or other accepted standards that promote commonality across DoD components" (Life Cycle Logistics)
Global business environment	"International Acquisition and Exportability (IA&E)" unit of competency (Program Management)
Business acumen (explicit reference)	"Business acumen" functional unit of competence in Engineering and PQM models, both of which include ten competencies that each relate to one or more of the key terms

SOURCE: RAND analysis of career-field competency models.

- Business-Cost Estimating
- Business-Financial Management
- Contracting
- Engineering
- Life Cycle Logistics
- Program Management
- PQM.

Conversely, career fields with lower Section 843–related knowledge needs based on our analysis include

- Facilities Engineering
- Information Technology
- Purchasing
- Science and Technology Management
- Test and Evaluation.

By comparing these designations with indicators of relative need, as expressed in DAWIA certification requirements and conveyed to us in our interviews, as we do below, we can gain greater insight into the extent to which they reflect actual relative need across the career fields.

DAWIA Certification Requirements

We used a two-step process to gauge the relative need for Section 843–related knowledge, as indicated by the DAWIA certification requirements for each career field. First, we examined DAU course names, concept cards, and course objectives to identify courses that convey business acumen, knowledge of industry operations, and/or knowledge of industry motivation.[31] Next, we reviewed DAWIA certification requirements for all levels of all career fields to determine which career fields require which DAU Section 843–related courses. The 19 DAU training courses that met both of these conditions—they convey the types of business-related knowledge in Section 843 and are *required* for certification for at least one career field—are listed in Table 2.2. The full

---

[31] Our list of Section 843–related DAU courses includes but is not limited to the set of courses on a briefing slide provided to us by an official at DAU entitled "DAU Core Courses Covering Understanding Industry Competencies" and those flagged in an unpublished 2013 DoD-sponsored study. As we describe in Appendix B, we developed a list that contains two levels of Section 843–related courses: a narrower list that includes those validated by at least one of these two outside sources and those that appear most directly related to a narrower conception of the three key terms and a broader list that includes the full set of courses that appear to convey knowledge related to broader definitions of the key terms. Our analysis included both DAU training courses and continuous learning modules (CLMs); however, for the purposes of the "need for knowledge" indicators developed in this chapter, we restrict the set of courses we considered to training courses that met our stricter standard for conveying Section 843–related knowledge (i.e., that are on our narrower list). Note that CLMs did not have course objectives available for our review, so our determinations were based on the course names and concept cards only.

Table 2.2
Section 843–Related DAU Training Courses Required for DAWIA Certification by at Least One Acquisition Workforce Career Field

Course Number	Course Name
ACQ 101	Fundamentals of Systems Acquisition Management
ACQ 202	Intermediate Systems Acquisition, Part A
ACQ 203	Intermediate Systems Acquisition, Part B
ACQ 315	Understanding Industry (Business Acumen)
BCF 110	Fundamentals of Business Financial Management
BCF 205	Contractor Business Strategies
CON 100	Shaping Smart Business Arrangements
CON 121	Contract Planning
CON 124	Contract Execution
CON 127	Contract Management
CON 170	Fundamentals of Cost and Price Analysis
CON 200	Business Decisions for Contracting
CON 270	Intermediate Cost and Price Analysis
CON 290	Contract Administration and Negotiation Techniques in a Supply Environment
CON 360	Contracting for Decision Makers
ENG 301	Leadership in Engineering Defense Systems
EVM 101	Fundamentals of Earned Value Management
LOG 235	Performance-Based Logistics
LOG 340	Life Cycle Product Support

SOURCE: RAND analysis of DAU course names, concept cards, and course objectives.

set of DAU courses that we identified as conveying this knowledge, including CLMs and training courses, and without regard to whether they are required, recommended, or absent from DAWIA certification requirements, is included in Appendix B.

For each career field, we counted how many of these 19 courses are required for DAWIA certification at any level. We then calculated an average across career fields.[32] Career fields that require an above-average number of Section 843–related

---

[32] The average number of Section 843–related courses (based on our determinations) required for DAWIA certification is 4.9, ranging from one to 11.

DAU training courses for DAWIA certification were designated as having a "higher" relative need for Section 843–related knowledge, while those that require a below-average number of these DAU training courses are placed in the "lower" relative need category. As shown in Table 2.3, career fields with higher needs for knowledge related to business acumen, industry operations, and industry motivation, as indicated by the DAU courses required to achieve DAWIA certification, are Business-Cost Estimating, Business-Financial Management, Contracting, Industrial Contract and Property Management, Life Cycle Logistics, and Program Management.

Consistency of the Approaches

Both approaches yielded the same result: All career fields require at least some knowledge related to business acumen, industry operations, or industry motivation, but some career fields have a greater need for this knowledge than others. Table 2.3 summarizes our findings, noting for each career field whether it has a higher or lower relative need for Section 843–related knowledge (compared with the average of all career fields), as indicated by its competency model and by its corresponding DAWIA certification requirements. For ten of the 13 career fields that we could compare, our determinations of relative need are consistent across the two sources, which makes us more confident that our results reflect the appropriate characterization of these needs.

Discussion and Limitations

Our methods for evaluating the documented needs for business acumen, knowledge of industry operations, and knowledge of industry motivation using existing DoD sources of requirements provide suggestive evidence of the degree of relative need for these knowledge types across the AWF on a career field–level basis.

Several limitations to our methods deserve mention. As we mentioned in Chapter One, the key types of knowledge upon which our study focuses—business acumen, industry operations, and industry motivation—lack clear definitions and can be interpreted differently by different people. This introduces an unavoidable degree of imprecision into our determinations of whether individual competency elements or DAU courses relate to one or more of these terms and means that the quantitative metrics that we derive from these determinations are necessarily approximations.[33]

With regard to the competency-model analysis, while we attempted to reduce the impact of differences across the models by assessing the breadth of Section 843–related knowledge needs rather than attempting a count of Section 843–related competency elements, our approach may nonetheless favor more-extensive models to the extent that these models cover more ground. It also creates an additional layer of ambiguity by introducing category tags that themselves are open to interpretation and that subsume

---

[33] The degree of imprecision surrounding our determinations and tabulations also drives our decision not to present the counts themselves in this report, preferring to focus on broad higher and lower relative need categories.

Table 2.3
Relative Need for Section 843–Related Knowledge by Career Field, as
Indicated by Competency Models and DAWIA Requirements

Career Field	Career Field Size	Evidence of Need in Competency Model	Evidence of Need in DAU-Required Courses
Business-Cost Estimating	1,434	Higher	Higher
Business-Financial Management	6,712	Higher	Higher
Contracting	30,748	Higher	Higher
Life Cycle Logistics	20,508	Higher	Higher
Program Management	17,727	Higher	Higher
Facilities Engineering	11,137	Lower	Lower
Information Technology	7,600	Lower	Lower
Purchasing	1,321	Lower	Lower
Science and Technology Management	3,977	Lower	Lower
Test and Evaluation	8,807	Lower	Lower
Engineering	*43,580*	*Higher*	*Lower*
PQM	*10,706*	*Higher*	*Lower*
Industrial Contract and Property Management	*391*	*Lower*	*Higher*
Auditing	4,209	N/A	N/A

SOURCES: HCI, 2018b; RAND analyses of career-field competency models and DAU course names, concept cards, and course objectives.

NOTES: "Higher" means that the career-field competency model incorporates an above-average number of categories of Section 843–related knowledge or the DAWIA certification requirements include an above-average number of DAU courses with significant business-related content using RAND criteria for this type of knowledge. "Lower" means a below-average number. Rows in italics indicate potential inconsistency in need expressed in the competency model and in the DAWIA requirements. N/A indicates that the career-field competency model and DAWIA requirements are not directly comparable with those for other career fields.

more-granular types of knowledge that may be included in the models but that are not explicitly called out in our tags. Moreover, in both the competency model and DAWIA requirement analyses, we do not attempt to assess the depth of Section 843–related knowledge needs—a model either covers a type of Section 843–related knowledge or it does not, and a course either conveys Section 843–related knowledge or it does not. Our methods also do not distinguish between differences in knowledge needs or required levels of proficiency within career fields for individuals in different positions or stages of their career.

Finally, it is important to keep in mind that this section describes Section 843–related knowledge needs as indicated by *existing* DoD sources. If there are knowledge needs that are not incorporated into the competency models or DAWIA requirements, those are absent from this analysis. Thus, the knowledge-need determinations from both approaches described herein should be considered in tandem with evidence from interviews and other sources that describe the nature of the need for business acumen and knowledge of industry and how such knowledge needs may vary across the acquisition workforce by career field or by other dimensions.

Interview Perspectives on Career-Field Knowledge Needs

Our interviews with DoD stakeholders and SMEs, including DACMs, functional leaders, HCI leadership, and DAU center directors, provided us with another valuable source of evidence to inform our assessment of the need of business acumen, knowledge of industry operations, and knowledge of industry motivation across the AWF career fields. The interviews often included a richness of detail that allowed us to supplement our broad characterizations with more specifics on which types of Section 843–related knowledge are needed in which career fields and why. Consistent with a core finding from our review of the competency models and DAWIA requirements, some interviewees underscored that all career fields need at least some level of the knowledge specified in Section 843. One stated, "I think to some degree probably everybody should have a little bit, it's how much do you need to know" (DoD 13), while another commented: "Anybody working for the government who ultimately relies on industry providing something for them, it's important that they have some understanding" (DoD 12).

Though all AWF career fields may have at least some need for business acumen and knowledge of industry, our interviews suggest that some career fields have a greater need for this knowledge than others.[34] Table 2.4 summarizes our interview findings, which align closely with those from our competency model and DAWIA requirement reviews (as indicated by the last column of the table). As the second column indicates,

---

[34] During all of our interviews with DoD personnel, we explored the extent to which different AWF career fields needed the types of knowledge cited in Section 843. For those with a DoD-wide perspective, such as DACMs and HCI leadership, we tried to collect viewpoints on all the career fields. For those with a narrower focus, such as functional leaders and DAU center directors, the discussions tended to include only evaluations of their own career field or ones that they knew best instead of commenting on the others. We did not ask external interviewees to comment on different AWF career-field needs, reasoning that representatives from commercial firms or external education providers would not fully understand the range of career fields that constitute the AWF. However, some external interviewees did mention specific career fields without our prompting, and we included such remarks in this thematic analysis.

Table 2.4
Perceived Need for Section 843–Related Knowledge by Career Field, as Indicated by Interviewees and Compared with Competency Models and DAWIA Requirements

Career Field	Evidence Strength	Interviewee Opinion	Table 2.3 Determinations Competency Model: Required Courses
Contracting	Strong	Very high need; highest in the AWF	Higher: Higher
Program Management	Strong	Very high need; highest in the AWF	Higher: Higher
Engineering	Moderate	High need; helps at the intersection of technology and business	Higher: Lower
Life Cycle Logistics	Moderate	High need; view of need has varied in recent years	Higher: Higher
Science and Technology Management	Moderate	Lower need	Lower: Lower
Test and Evaluation	Moderate	Lower need	Lower: Lower
Business-Cost Estimating	Limited/mixed	Unclear—Some felt that the need was not that high, given less direct interaction with industry and DoD-specific nature of work	Higher: Higher
Business-Financial Management	Limited/mixed	Unclear—Some felt that the need was not that high, given less direct interaction with industry and DoD-specific nature of work	Higher: Higher

SOURCES: 2018 RAND Section 843 study interviews; HCI, 2018b; RAND analyses of career-field competency models and DAU course names, concept cards, and course objectives.

NOTES: Interview evidence strength is based on frequency of mention, richness of the discussion (e.g., explanation offered for opinion), and consistency of opinion across interviewees. Little to no evidence was available for the remaining career fields: Auditing, Facilities Engineering, Industrial Contract and Property Management, Information Technology, Purchasing, and PQM.

the strength of the evidence to support our characterizations of interviewees' opinions varied by career field. For some career fields, the evidence was strong, meaning that a relatively high share of interviewees mentioned the career field, statements about knowledge needs tended to include explanations, and views were consistent across interviewees. In other cases, the evidence strength was more moderate, indicating that, while views were consistent across interviewees who cited the career field, the career fields were not mentioned as frequently overall. For the last two career fields listed in the table, Business-Cost Estimating and Business-Financial Management, interviewee opinions were both limited in number and mixed in direction. For the remaining career fields, there was insufficient interview evidence to make a characterization about the need for Section 843–related knowledge at all.

Career Fields with Strong Evidence of Very High Section 843 Knowledge Needs

Interviewees consistently indicated that members of the Program Management and Contracting career fields have a very high need for the types of business-related knowledge specified in Section 843.

Interviewees explained that program managers have a particularly high need for this knowledge because of their high level of interaction with industry, responsibility for major programs, involvement in negotiations, and leadership role in the acquisition environment:

> Certainly I would say Program Management, number one. I mean, they're the people that are typically held responsible or accountable for the performance of the program. (DoD 21)

> Well, if you're a Program Manager a lot of times you've got all of these career fields under you. So having the kind of top-level corporate understanding of everything and how to make all the functions under you work properly is really important. (DoD 1)

> I think in actually relating with industry on a daily basis with regards to business acumen, yeah. I mean, engineers deal with their contractors and building and deploying and designing whatever system they have, but I guess really the Program Managers, they're [the ones] having that business acumen discussions about where are we on cost, where are we on schedule, are we going to get all of the capability within our current budget? Those types of things. (DoD 26)

On occasion, Contracting was mentioned ahead of Program Management in terms of need for industry-related knowledge, with one interviewee commenting: "[I]n Contracting . . . [it] probably [has] the most direct face with industry. . . . So, I think the Contracting perspective may be the most—if I dare say, the most relevant compared to some of the other career fields" (DoD 12).

Interviewees homed in on Contracting personnel's role in structuring contract incentives and their need to understand what motivates industry in order to make good business deals for the government:

> I'll add to Contracting that if a Contracting Officer understands what incentivizes industry, how small changes in the contract that don't matter much to us might impact industry's bottom line significantly. Knowing that those tradeoffs exist helps them come to a better business deal for us or more likely that they can reach a good business deal together. (DoD 1)

> If you're dealing with contracts, you might want to understand industry's perspective, as well, even financial management. (DoD 19)

[B]ecause we tend to stovepipe the negotiations of contracts, the contracting community also needs to understand the importance at least on the motivation part. (DoD 5)

In sum, Program Management and Contracting were viewed as the career fields with the greatest need for knowledge related to business acumen, industry operations, and industry motivation because of the role that members of these career fields play in interacting with industry, negotiating contract terms, and, particularly in the case of Program Managers, overseeing the acquisition process and having ultimate responsibility for ensuring that capabilities are acquired on time and on budget.

Section 843 Knowledge Needs Across Other Career Fields

For two career fields, Life Cycle Logistics and Engineering, the interviews provided moderate evidence that there is a relatively high need for Section 843–related knowledge; for two career fields, Science and Technology Management and Test and Evaluation, they provided moderate evidence that there is a relatively low need for this knowledge; for the Business-Cost Estimating and Business-Financial Management career fields, the interview evidence was mixed; and for the remaining six career fields, there was insufficient evidence to draw a firm conclusion.

Moderate Evidence of High Need

Some interviewees mentioned that members of the Life Cycle Logistics and Engineering career fields need business acumen and knowledge of industry, though opinions were not as strong, consistent, or frequently expressed as those on Program Management and Contracting. The need for Section 843–related knowledge among those in Life Cycle Logistics is an extension of the need for this knowledge in Contracting, according to a few people with whom we spoke, with logisticians taking on greater responsibility for setting requirements and tracking delivery as programs enter the sustainment phase:

> Typically the program matures, a lot of the contracting activity goes toward logistics and sustainment, eventually, and they need to have a keen sense of knowledge of what the interlaying factors are there. (DoD 21)

> The dicey part is we really need our Life Cycle Logisticians to be knowledgeable on that world of sustaining weapon systems but be smart enough—and they don't need to replicate what a Contracting Officer needs to be able to do but they need to understand enough so that they can speak intelligently to a Contracting Officer. And so much of that means understanding the nature of the risk that our work brings to a vendor and what the vendor can take on and what we can take on. (DoD 15)

Other interviewees noted that the perceived need for business acumen and knowledge of industry in Life Cycle Logistics waned before reasserting itself in recent years:

> [F]or a long time, there was a model that maybe our Logistics community would benefit from bringing in more folks that understood Walmart's model for logistics. Then a few years later, it seemed like that had been discussed a lot less or maybe some people even said that might not have been the right way to go. But I think we find that's swinging back the other way now and we're looking to leverage those best practices. (DoD 1)

> [I]t's probably maybe four or five years ago now, certainly it's been some years—that our community realized, hey, we have a perceived gap within our workforce for Life Cycle Logistics in terms of that business acumen, understanding how industry operates, understanding what some of the key considerations are in that realm, what are some of the motivations for industry. (DoD 8)

With regard to Engineering, Section 843–related knowledge needs were perceived as relatively high among interviewees who offered an opinion about this career field, though the rationale underlying the need for this knowledge in Engineering differed from the explanations given for the other three career fields perceived by interviewees as having a high need. Specifically, the need for business acumen and industry-related knowledge in Engineering was expressed as a need to be able to operate at the intersection of technology and business, to support those in more directly industry-facing acquisition roles, and to inform the development of business deals that result in the delivery of technologically advanced systems at a fair price and on a reasonable timeline:

> A lot of our cost estimating is not necessarily done by the cost estimating community. In many program offices, it's the engineering community that is doing that. So I would put those two together with regard to the financial piece of how do you estimate how much it's going to cost to create, design and build System X and then understanding the technology piece, the engineering and the science and technology management piece has to really stay engaged with our industry partners to see where we are with regard to technology readiness levels. (DoD 5)

> But if you don't have a PM with a technical background, they have to be fairly smart and then you have to have business-savvy engineers—at least a portion of them—working under your program. So they know how to do effective contracting and how to talk to financial managers and how to deal with obligations and expenses, etc., etc. So, I think it's very important but it's very important for engineers to be very highly technically qualified when they come in, so the government can be the smart customer. (DoD 18)

Moderate Evidence of Lower Need

Interviewees were more likely to discuss the career fields that they believed to have a high need for Section 843–related knowledge than to cite those with a lower need, but we noted moderate evidence of lower need for two career fields: Science and Technology Management and Test and Evaluation. This is not to say that these career fields have no need for Section 843–related knowledge (note, for example, that Science and Technology Management was cited alongside Engineering above as an example of a field that needs to operate at the intersection of business and technology). Rather, the perceived need is lower relative to other career fields, in large part due to the more inward-focused nature of most positions in these career fields:

> Well, I'm sure it's on a scale. Something like Science and Technology, they don't always have the direct face to industry like we have. (DoD 12)

> Science and Technology Management, you know, I guess the senior people in that organization would but . . . the people that are below the very senior-most person, they're typically dealing with the technicalities of system performance versus really caring about what industry cost and pricing is about. (DoD 21)

> [J]ust don't see the testers needing a whole lot of business acumen, knowledge of business operations or motivation in the test evaluation world. They're very, very insular in their processes and way of doing things. (DoD 10)

Mixed Evidence

The two business career fields—Business-Cost Estimating and Business-Financial Management—were mentioned by fewer interviewees, were typically discussed in tandem, and elicited mixed opinions. Of those who did mention these career fields, some thought that they have among the greatest needs for Section 843–related knowledge; for example, one interviewee commented that "Business-Cost Estimating and Financial Management is pretty high up there" (DoD 3), while another cited the OPM definition of business acumen in expressing a view that there is a high need for that type of knowledge in these career fields: "I think we just kind of thought that the cost-estimating and the financial part of it, as it related back to the definition of business acumen that OPM defined, is kind of where we made that connection" (DoD 1).

Other interviewees provided caveats in describing the need for business acumen and knowledge of industry among members of the two business career fields. One noted that "they don't do as much interaction with the industry folks as the Program Managers and the Contracting Officers" (DoD 26). Two others cited the DoD-unique aspects of the work as weighing against their need for knowledge of industry:

> I think they do [need business acumen and industry knowledge] and I think we try to teach some of that stuff, of looking at profit margins and, you know, basic things like that, that drive industry behavior. But it's kind of from our perspective

rather than sitting there on the industry side trying to figure out how to make a buck. (DoD 23)

[F]rom [a Financial Management] perspective, for instance, we do things uniquely in DoD given the [Planning, Programming, Budgeting, and Execution] process that we're in, that is unique from any other federal agency. (DoD 19)

On balance, we did not believe that there was sufficient interviewee evidence to characterize the perceived level of need for Section 843–related knowledge in these two career fields. However, our preceding analyses of the career-field competency models and DAU-required courses for DAWIA certifications both suggested a higher-than-average need for knowledge related to business acumen, industry operations, and industry motivation for these two career fields.

Insufficient Evidence

For the remaining six career fields, while the evidence from the interviews was not necessarily mixed, there simply was not enough of it for us to draw conclusions about an evidence theme. These career fields were Auditing, Facilities Engineering, Industrial Contract and Property Management, Information Technology, Purchasing, and PQM. The lack of mentions itself may be an indicator of lesser perceived need for Section 843–related knowledge, or it could reflect the comparatively smaller size of several of these career fields, making them less salient to our interviewees during our discussions.

Variation in Knowledge Needs by Experience Level

In the previous sections, we described the variation in Section 843–related knowledge needs across the acquisition workforce career fields, as indicated by existing requirements and as expressed by the DoD stakeholders that we interviewed. Our career field–focused approach was driven by our initial hypothesis that Section 843–related knowledge needs likely varied by career field, and it is consistent with how T&D for members of the AWF is designed and implemented. However, a limitation of this approach is that it masks variation within the career fields, most notably differences in needs by experience level.

The sources we reviewed and the interviews we conducted consistently suggested that Section 843–related knowledge needs are higher for more-experienced personnel—people who already have their DAWIA Level III certification, who may be in CAP or KLP positions, and who are more likely than more-junior personnel to be leading programs and interacting directly with industry and other outside stakeholders. To that end, we reiterate that "business acumen" is explicitly listed among

OPM's ECQs to be in the SES[35] and among the core competencies in the DoD Civilian Leader Development Framework included in DoDI 1430.16.[36]

Moreover, while most career-field competency models do not parse the competencies into those required for individuals at different points in their careers,[37] we understand from our targeted discussions with DoD officials that, in practice, the expected level of competency often varies by level of seniority, and, in some cases, certain competencies in a given career-field model may not apply at all, depending on an individual's precise role. Based on our interviewees' opinions of how needs for business acumen and knowledge of industry vary by experience level, we expect that competencies that relate to one or more of the Section 843 knowledge types are among those that apply differently to junior- versus senior-level personnel. Interview comments consistent with such a view included the following:

> [A]t the very junior levels, the entry-level folks coming into our career field probably don't need to know a whole lot [of Section 843–related knowledge], in my opinion. They need some broad exposure. . . . [I]t's more the middle- and the upper-level folks within our community; more experienced, more seasoned, more leadership level that, as you move up in authority, you're going to have to have a much more broadly based understanding. And certainly as you get up to I'll say Level III–certified workforce members and moving into product support manager positions, you absolutely have to have that understanding. (DoD 8)

> I would say certainly the senior engineer from the Engineering Department, senior contracting representative or contracting officer. Typically the product support manager, who's the senior logistician on a program. . . . But I think, that said, the program manager, the senior engineer, the senior logistician, certainly the senior contracting individual, those are the key people. . . . Typically Level 3. You might get some exposure to the terminology a little bit in Level 1 and 2 but you really need to have a little bit more scar tissue, I think, to have a full appreciation of the various aspects. (DoD 21)

Similarly, while we analyzed Section 843–related knowledge needs as embedded in the DAWIA certification requirements for career fields in the aggregate to simplify the presentation (i.e., we did not parse the DAU courses into those required for each of the three levels of certification), we recognize that these certification requirements also point to differences in knowledge needs for Level I, Level II, and Level III personnel.

---

[35] OPM, undated.

[36] DoD, *Growing Civilian Leaders*, DoDI 1430.16, November 19, 2009.

[37] As noted previously, Contracting has a competency model with five proficiency level standards corresponding to each of 52 "technical" competency elements (but not for its "non-acquisition-unique" competencies), while Program Management has a model with basic, intermediate, and advanced descriptions for all competency elements.

For example, the DAU course that most clearly conveys knowledge related to industry and business acumen (ACQ 315, "Understanding Industry—Business Acumen") is a requirement for Level III Program Managers (and is an option to fulfill a Level III requirement for Contracting and Life Cycle Logistics) but is not required for lower levels of certification.

In sum, while the bulk of our analysis has explored differences in the need for business acumen, knowledge of industry operations, and knowledge of industry motivation across the AWF career fields, we recognize that there are differences in needs within the career fields and that these differences may be just as, or more, pronounced.

Summary

AWF leaders have implemented the competency management framework adopted by DoD, but, as we discussed in this chapter, the resulting competency models vary in structure across the acquisition career fields, in particular in how the competencies included therein are mapped to experience levels and proficiency levels. This variation, as well as the lack of formal definitions of the terms *business acumen*, *industry operations*, and *industry motivation*, complicates the task of assessing how the various competency models express the need for the types of knowledge in Section 843.

The analysis in this chapter does not attempt to determine conclusively or validate which acquisition career fields have a need for business acumen, knowledge of industry operations, or knowledge of industry motivation. However, it does show that, using our definitions of these three types of knowledge, AWF competency models and DAWIA certification requirements are fairly consistent in the expression of the relative need of different career fields for these types of knowledge, and some career fields appear to have a higher relative need for this knowledge than others. For the two career fields, Program Management and Contracting, there was strong evidence in the interviews that the need for Section 843–related knowledge was very high, which corroborates findings from our competency model and DAWIA certification analyses that these career fields have a higher relative need for this knowledge. This also supports our hypothesis that the relative need for business acumen, knowledge of industry operations, and knowledge of industry motivation varies by career field, which may affect the demand for resources to close gaps that exist in these areas of knowledge.

CHAPTER THREE

Gaps in Knowledge Related to Business Acumen, Industry Operations, and Industry Motivation

In this chapter, we identify knowledge gaps related to the three types of knowledge referred to in FY 2018 NDAA Section 843—business acumen, industry operations, and industry motivation—that appear to exist within the AWF. We provided definitions of those terms earlier in the report and repeat them here for ease of reference:

- Business acumen: In addition to the ability to manage human, financial, and information resources strategically (OPM definition), business acumen is an understanding of industry behavior and trends that enables one to shape smart business decisions for the government.
- Industry operations: This includes plans and procedures used within an industry to provide a product or service. The need for knowledge of specific practices may vary depending on an employee's contribution to the acquisition mission. Some industry operations may be business oriented, while others may be at the confluence of business and technical knowledge—i.e., "techno-business."
- Industry motivation: This includes the range of considerations and motivations that factor into the decisionmaking of organizations in industry, including profit and revenue, market share, management and employee incentives, shareholder considerations, perspectives on risk, and the need to maintain position in a competitive environment. The relative weights of these factors may vary by industry and over time.

First, we describe what we learned regarding how DoD determines which knowledge gaps are present in the AWF. Following that, we review our approach to identifying the knowledge gaps pertinent to this study and the limitations of that approach. We then discuss specific gaps related to Section 843's focus, including the evidence that supports our conclusions.

DoD Approach to Identifying Knowledge Gaps

At the outset of our research, we looked for policy or other guidance on the gap assessment process that DoD should use to identify knowledge gaps.[1] We noted that both Vols. 250 and 410 of DoDI 1400.25, which pertain to the DoD civilian personnel management system, identify organizations and individuals with responsibilities related to competency gap assessment, and DODI 1400.25, Vol. 250, repeatedly mentions the Defense Competency Assessment Tool (DCAT), a tool intended to replace component-specific tools used to assess individual and workforce-level occupational competency gaps. In the 2014–2015 time frame, the Defense Civilian Personnel Advisory Service (DCPAS) began applying the DCAT to mission-critical occupations[2] and issued two Defense Competency Assessment Tool Implementation Guides, one for supervisors and one for employees, to facilitate its use.[3]

Although the DCAT and pertinent guidance have been available for several years, use of the DCAT was not discussed during our DoD interviews. Instead, we heard about a fragmented approach to knowledge (or competency) gap assessment. First, rather than a centralized approach to gap assessment, we found that many organizations are involved in DoD efforts to identify and address gaps. When FIPTs develop and revise competency models, some of their assessments of competencies needed will include some consideration of how proficient AWF personnel working in a specific career field currently are in those competencies. In some cases, individual military services or agencies assess gaps in their own acquisition workforce, and we also learned about efforts to identify and address gaps at the command or sub-agency level. For example, the Navy has "national leads"—individual career-field managers responsible for determining the nature and extent of knowledge gaps within their segment of the Navy acquisition workforce.

Not only do the organizations that may conduct a gap assessment and the focus of such an assessment (full AWF or a portion) vary, but the methods used do as well. During our interviews with DoD SMEs, we asked how they became aware of knowledge gaps or what evidence they collected to indicate that gaps existed. The responses ranged from "I don't know if we've really had any kind of determination of what that

---

[1]   Although we recognize that there are differences between knowledge, competencies, and skills, for the analysis in this chapter, we looked for evidence of assessment processes and gaps related to all three indicators of AWF cognitive capabilities. We followed this approach in part because of the heavy emphasis on competencies within AWF training and development and also because competencies may include a mix of knowledge, skills, and abilities. Specifically, a competency is "[a]n observable, measurable pattern of knowledge, abilities, skills, and other characteristics that individuals need to perform work roles or occupational functions successfully" (DoDI 1400.25, Vol. 250, 2016a, p. 21).

[2]   DCPAS, *Defense Competency Assessment Tool (DCAT) Frequently Asked Questions (General)*, January 2015.

[3]   DCPAS, *Memorandum for Department of Defense Civilian Employees and Supervisors in Mission Critical Occupations: Follow-up on the Defense Competency Assessment Tool, Initial Operating Capability*, October 8, 2014.

gap [in knowledge types cited in Section 843] is and what aspects of business operations our AWF doesn't know" (DoD 13) to descriptions of using research methods, such as surveys, to identify gaps. Table 3.1 includes examples of the evidence provided by our interviewees that conveys the range of methods DoD uses to assess knowledge gaps. We differentiate between responses that indicate more-deliberate strategies and the use of scientific research practices, such as surveys, and those that described less deliberate or scientific means.

Overall, we found that DoD's approach to assessing knowledge gaps at the time of our study was very decentralized, with variation in how and when gaps are assessed. It should be noted, however, that DoD is not unique in this regard: GAO stated in its 2015 and 2017 "High Risk Series" reports that OPM, the Chief Human Capital Officers (CHCO) Council, and federal agencies needed to improve their efforts to identify and address critical skill gaps. Specifically, GAO stated, as of 2015, that these organizations "needed to do additional work to more fully use workforce analytics to identify their gaps, implement specific strategies to address these gaps, and evaluate

Table 3.1
Evidence from Interviewees Regarding Approaches to Identifying Knowledge Gaps

Less-Scientific or Standard Approaches	More-Scientific Approaches
I have to admit, it's largely discussions. (DoD 15) We have tried different things and we haven't settled on any one particular model that is like, "Do this or fill this out and we'll know where your gaps are." (DoD 14) We have representation from the secretariat side, the Assistant Secretary of the Navy for Acquisition, ASN RDA folks, but we also have representation from the fleet side [similar for] the Air Force . . . the Army. That's still good, but that's not enough. You also have within our community the System Commands or the Major Commands underneath those Headquarters element. So they all are represented. They all know intuitively the workforce professional development requirements and expertise required of their own communities and they are going to feed that information right back into the Functional IPT process. They're all represented or all play in the competency review process. So every time we do a review of our competency set, they're going to be very vocal to tell OSD, "Hey, we have a gap, we have a knowledge gap in our workforce and we need more focus on these kind of things." (DoD 8)	We've done studies and surveys and had [career field] forums on a regular basis that get feedback from [career field personnel] on where their training and experience gaps are, as well as the people on their staff. (DoD 16) We did a competency assessment . . . where we took the competency model, put it out across the whole career field. . . . Based on their inputs, we looked through their assessment. We found some areas . . . where people weren't as proficient as they should be. (DoD 4) The DoD AWQI Workbook is the source document for functional/technical competencies. Functional/technical competencies in the AWQI e-workbook are distinct for each career field and vary in number. . . . To facilitate a positive user experience, we limited the number of functional/technical competencies to 18 or less, as determined by [service-level functional leaders] for their respective [career fields]. As a result, the functional/technical competencies for assessment models are unique to each [career field]. However, all assessment models share the same development, deployment, and analysis process. Hybrid terms of reference from the AWQI e-workbook and the National Institutes of Health Competency Proficiency Scale were combined to describe the functional/technical proficiency scale. We conducted the assessment . . . over the course of three months to not overwhelm the entire . . . population. (DoD 3)

SOURCE: 2018 RAND Section 843 study interviews.

the results of actions taken so as to demonstrate progress in closing the gaps."[4] In 2017, GAO reported improvements to the analytical methods used to identify gaps but indicated that more progress was needed.[5] In FY 2018, HCI launched a five-year effort in collaboration with DCPAS to use the DCAT for career field–level technical competency assessments. The process HCI developed is depicted in Figure 3.1. The process is estimated to take about 11 months per career field and includes the following stages:

1. a pre-planning stage to determine the parameters of the effort and to populate career-field data elements
2. a stage (Phase I) to conduct substantive background research via a literature review and SME panels, which results in a list of competencies
3. a stage to validate the model and assess competency gaps using DCAT (Phase II)
4. a stage for a post-DCAT validation using a SME panel and to prepare the final report.[6]

The full effort is slated to run from FY 2018 through FY 2022. The Lifecycle Logistics career field conducted its DCAT-based survey (Phase II) in spring 2018 and convened a post-DCAT panel in late July. Purchasing and PQM efforts were in the Phase I stage at the time of this report's publication.[7]

The relationship between this effort and the FIPT process for developing career-field competencies is unclear. We mentioned in Chapter Two that the DoD competency model approach has five tiers of competencies: The highest level is Tier 1 (core

Figure 3.1
Competency Gaps Assessment Process Using DCAT

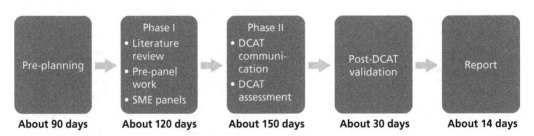

SOURCE: Based on HCI, "Competency Assessment Overview," presentation slide, undated.

---

[4]   GAO, *High-Risk Series: Progress on Many High-Risk Areas, While Substantial Efforts Needed on Others*, GAO-17-317 High-Risk Series, 2017, p. 61.

[5]   GAO, 2017, p. 63.

[6]   HCI, *Competency Assessment Overview*, presentation, undated(a), provided October 5, 2018; and HCI, *Competency Management Process*, presentation, undated(b), provided by OUSD(A&S) on October 5, 2018.

[7]   HCI, *A&S Human Capital Initiatives (HCI) Updates Presented to Workforce Management Group (WMG)*, September 5, 2018d, briefing provided to RAND study team by WMG group member on October 15, 2018.

competencies), and Tier 2 is primary occupational competencies.[8] Functional leaders typically write their competencies at Tier 2,[9] but for the DCAT competency gap assessment process, we learned through our targeted discussions that the competency level included in the DCAT approach is higher than those included in the competency models. However, we also were advised that the competency lists developed through the DCAT-based approach exclude what the DCAT team considers to be Tier 1 "soft" skills—some of which we found were included in FIPT competency models.[10] Finally, it is also unclear whether members of the FIPTs are invited to participate in the DCAT SME panels or whether the DCAT assessments will provide feedback for the revision or validation of FIPT-produced competency models.[11]

Although the plan to apply the DCAT methodology to all acquisition career fields over a five-year period is promising, it is unclear which DCAT assessments will include the three types of knowledge on which Section 843 focuses. To date, DoD efforts to assess gaps have not consistently looked at business acumen, knowledge of industry operations, and knowledge of industry motivation, and they may not in the future, given DoD's tendency to focus on technical competencies at this level. Moreover, as mentioned in Chapter One, shared, formal definitions of these types of knowledge are lacking. In addition, the extent to which competency models include them varies by career field, and their inclusion often is not explicit (e.g., a model that does not refer expressly to "industry operations"). Figure 3.2 shows the results of our analysis of competency elements included in competency models and AWQI workbooks for use of language that we determined was related to the three types of knowledge specified in Section 843. As explained in Chapter Two, we developed a set of keywords based on our definitions of business acumen, industry operations, and industry motivation–related knowledge and pertinent discussions in our interviews and applied those keywords to competency elements included in career-field competency models. The figure summarizes how many models include elements related to each of the keywords that we regarded as proxies for knowledge related to business acumen, industry operations, and industry motivation. For example, all of the models include competencies related to technology/technical management and human capital management; conversely, only three models include the explicit words "business acumen." None of the models included the specific phrases "industry operations" or "industry motivation." Finally,

---

[8]   DoD, DoDI 1400.25, Vol. 250, 2016a.

[9]   DoD, DoDI 5000.66, 2017a.

[10]   For example, in the October 17, 2018, targeted discussion with DCPAS, "communications" was cited as a Tier 1 core competency that would not be assessed using DCAT. Communication is included as a professional skill in the PQM career-field competency model.

[11]   Because they are still in the earliest stages of this effort, DCPAS declined to share examples of its competency models (October 22, 2018, e-mail from DCPAS), so we were unable to compare them with the FIPT competency models we received or with competencies included in the AWQI e-workbooks.

Figure 3.2
Tally of Section 843–Related Competency Elements Across Career-Field Competency Models

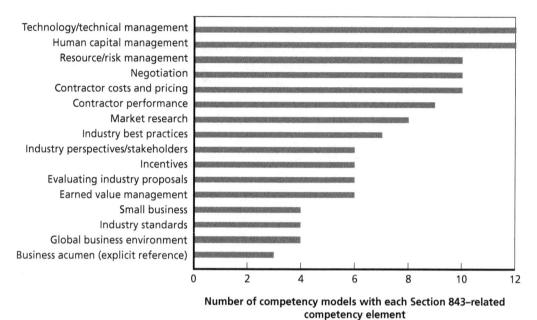

Number of competency models with each Section 843–related competency element

NOTE: Thirteen career fields are included in this analysis of 12 competency models: Business-Cost Estimating and Business-Financial Management were combined. Auditing was not included in the analysis because the career field did not have a comparable model in the AWQI e-workbooks.

most of the competency models did not include desired proficiency levels for the competencies therein. For example, while all the models included a human capital–related competency element, most did not indicate whether level 3 (intermediate) proficiency was required or whether a different proficiency standard applied.

Together, these observations suggest that clear and consistent proficiency standards for these three types of knowledge are not available, which limits DoD's—and our—ability to conduct a precise gap assessment.

RAND's Approach to Identifying Knowledge Gaps in the AWF

We used a multifaceted approach that included several data sources: competency models and AWQI e-workbooks; our interviews with DoD and industry SMEs; and publications that referred to knowledge gaps for the AWF, specific acquisition career fields, or specific acquisition occupations. We reviewed competency models and AWQI e-workbooks for evidence of required proficiency levels related to business acumen, knowledge of industry operations, and knowledge of industry motivation; these standards would be the "yardstick" against which to gauge proficiency levels. The interviews included participants' views of knowledge gaps and references to additional evidence sources, and the publications covered knowledge gaps in varying levels of detail, generally relying on interviews or survey data to support their conclusions. Most of the publications we reviewed focused on the AWF as a whole. We also had the opportunity to review service-level gap assessments not publicly released, as well as one career field–level assessment, DoD's 2014 *Study of Program Manager Training and Experience*.[12,13] We limited our publications review to the last five years, the rationale being that gaps cited earlier than that time frame likely were closed, and if they were not, they would be mentioned in more-recent publications as well.

In this chapter, we focus on the most salient and critical gaps in business acumen, industry knowledge of operations, and knowledge of industry motivation—i.e., those that were cited in multiple sources of evidence or extensively detailed in one. The lack of career field–level required proficiency ratings for the three types of knowledge cited in Section 843, coupled with the paucity of career field–level gap assessments, precluded us from systematically reporting gaps at the career-field level or estimating the size of gaps in terms of the number of personnel affected. In addition, although we focused on recent publications—those released within the past five years—it is possible that progress has been made on closing some of those gaps, such that older estimates of gap magnitude would no longer be valid.

Gaps Related to Business Acumen

Business acumen was discussed less extensively in the interviews than the other types of knowledge referred to in Section 843. We presented the three Section 843 types of knowledge together in our questions, and while no one expressly stated that gaps in business acumen did not exist, people tended to focus on gaps related to knowledge of industry operations or industry motivation. Some interviewees[14] claimed that there were business acumen gaps, but their descriptions of such gaps were more closely aligned with our definitions of industry operations or industry motivation. For example, one interviewee told us, "I have to double-check the exact time—it's probably

---

[12] We reviewed the full report, which is not cleared for public release, but in this report we provide only details publicly available in K. W. O'Donnell, "A Meeting of the Minds: Expanding Training and Understanding Between Industry and Government," *Defense AT&L*, January–February 2018, pp. 2–7.

[13] The Engineering career-field leaders provided us with the *Technical Leadership Development Guidebook* prepared by Stevens Institute of Technology Systems Engineering Research Center in 2016. We reviewed this report carefully and found that it covered potential proficiency standards, as well as a proposed way to assess the level of competency attainment, but it did not include the results of such an assessment.

[14] As we noted in our methodology discussion, to identify themes from interviews, we considered not only how often a topic was cited within and across interviews but also the richness of the discussion and the level of agreement across interviewees regarding a specific topic or theme. Accordingly, we do not provide interview counts in the report and instead use the term "some" when describing a theme present across multiple interviews.

maybe four or five years ago now, certainly it's been some years—that our community realized, hey, we have a perceived gap within our workforce . . . in terms of business acumen: understanding how industry operates, understanding what some of the key considerations are in that realm, what are some of the motivations for industry" (DoD 8). We found something similar in one of the unpublished DoD studies: The research team cited gaps that it regarded as aspects of business acumen, but many of the knowledge elements pertained to understanding various aspects of how private companies function and to familiarity with industry motivation and incentives (e.g., "key corporate motivations that ensure meeting financial objectives and reliable cash flow, profit, and growth opportunities").

A small number of interviewees discussed business acumen–related knowledge gaps in general terms such as the following:

> The business acumen is understanding how to deal with a business situation and then as it relates to the Acquisition process that all lead to good outcomes. Yeah, so our business acumen . . . I mean, I think we have a fairly significant challenge within the [service] with regard to business acumen and I think that we need to do more in the development of business acumen. We've done lots of things to try to figure out how we can improve on our business acumen, but there's been a big focus in the [service] on technical competence and sometimes at the expense of the business acumen piece of it. (DoD 3)

Still others focused on specific aspects of business acumen related to an acquisition professional's ability to manage financial resources effectively: risk management and earned value management (EVM). For example, during his discussion of business acumen, one interviewee told us:

> So I think a gap right now that we've identified as something that we need to have as one of our competencies [is] risk management. . . . I can't get away from the idea now that I think we are not as comfortable as [specific AWF profession] dealing with risk as we need to be. And, I don't know, that gap may exist not just inside government [specific AWF career field], but that's probably inside our community, our engineering community, industry, and government. (DoD 14)

DoD-sponsored studies, both published and unpublished, were also suggestive of risk management–related deficiencies in the AWF. In one of the unpublished studies that we reviewed, competency development efforts to improve proficiency in risk management were recommended for three career fields. In addition, in the DoD 2014 *Study of Program Manager Training and Experience*, 44 percent of those in the Program Management career field either responded "no" or "unsure" when asked whether acquisition training was sufficiently practical and comprehensive to enable them to manage or deal effectively with managing risk and opportunity. This may be an area in which some improvements are in progress, however. For example, in 2009, the response to the same question was 7 percentage points higher (i.e., 51 percent of program managers were negative or unsure about the sufficiency of acquisition training related to risk management).[15]

EVM was cited repeatedly in interviews and studies as an additional area of concern in the business acumen knowledge domain. As one DoD leader put it:

> [B]usiness acumen isn't specific to one or two or three of those career fields. It's just a general competency or awareness that all of the workforce needs—especially at an executive level or a senior level, when people are in positions or authority and responsibility and accountability and making those type of decisions. So it's trying to understand how we manage those cross-cutting things. Earned value management is an example. It's not really a career field but it's a competency that is needed across different career fields. And so we're trying to identify now—put some training in place and trying to identify what workforce members, what positions really need that training. (DoD 26)

The studies we reviewed imply that the need for EVM and the extent of a gap may vary by career field. For example, in an unpublished DoD study conducted in 2017, one career field was perceived as deficient in EVM performance analysis and management. Conversely, program managers reported marked improvement in EVM in recent years. Specifically, in the DoD 2009 *Study of Program Manager Training and Experience*, only 37 percent of program managers felt that acquisition training was sufficient to enable them to perform or use earned value, and in the 2014 update, that figure increased to 81 percent.[16]

Gaps Related to Knowledge of Industry Operations

> I think it is important . . . to have an appreciation for how business operates, and it varies by sector. If you're in the construction portion of DoD, the business acumen around that market is different than it is professional services or manufacturing. But time and time again, we see that there is really no appreciation for how business operates, the role of time and money in that activity, and that's a shame. That puts the Department at a disadvantage, particularly when they're contracting with the private sector for providing goods and services. (Industry 2)

Both DoD and industry interviewees discussed at length the need for the AWF to have knowledge of industry operations and referred to broad industry operations–related gaps in remarks such as the one above as well as to more specific gaps related to different aspects of industry operations. Table 3.2 provides exemplary quotes pertaining to each

---

[15]   Study results cited in O'Donnell, 2018, pp. 1–7.

[16]   O'Donnell, 2018.

Table 3.2
Interview Evidence Regarding Industry Operations Knowledge Gaps

Type of Knowledge Gap	Exemplary Quotes
Financial aspects	"So one [gap] would be finances, how industry finances its operations, how it looks at cash flow, how it looks at overheads and direct cost. Those kind of things and pretty much the financial side of industry and how it operates is pretty important. Probably another one is just essentially the whole area around industry structure and how impactful the structure and operations are to the business deal." (Industry 7)
	"I started reading analysis from outside consultants that we pay for that started talking about their stock price, what their internal business goals were, what's driving their stock price, what pressures they're under, and, most importantly, what factors they considered most important in negotiations, whether it be cash flow, whether it be timing, whether it be incentives, etc., etc. And that information is not making it out to the field and that's what I call business intelligence. So, that's what I consider to be stuff that goes beyond the normal reading of, you know, USA Today or whatever. That type of analysis is what I think the field is lacking, along with a couple other things . . . " (DoD 25)
	"Some of the understanding of a corporation portfolio management, why [do] they do mergers, acquisitions and divestitures, and what are the associated financial implications for the government? For example, when Northrop Grumman spun off Huntington Ingalls, was the right amount of debt-to-equity and retained earnings shifted with the new spinoff that was now going to be making ships for the federal government? Were they as liquid as they were before they were spun off? And what do you look for in the 10Ks or the annual reports that signal where an industry is going to go? And what levers can you pull as a senior executive that influence where they go or ensure that the needs of the defense of the nation are taken care of? That's a gap and it is a tough [one]." (DoD 9)
Supply chain management	"We could do much better with understanding suppliers and supply chain management. The majority of that is [now] much more focused on how we do that internal to the department." (DoD 5)
	"I would say that the biggest knowledge gaps that I see today in the government are an in-depth knowledge of industry cost pricing, G&A [general and administrative] overhead, profit and subcontractor management, or supply chain management." (DoD 21)
Small business	"So really understanding the nuances of how the businesses make money would be helpful and I suspect most small business folks and maybe acquisition at large don't have real depth of knowledge there. Recognizing the burden of government compliance. You know, in big, large companies they'll have a compliance officer and a staff to deal with all that, but with a small company they're wearing multiple hats. So that's a lot of burden, and I don't know that on the government side we appreciate the impact of regulation on a company. So, that would be helpful. In Contracts, they're going to just throw a bunch of FAR clauses on them not even thinking. And if they understood the implications of each one there may be some that they would say, 'Well, you know, we really don't need this,' or 'If we do put this on, it's reasonable that their rates are going to be impacted in some way,' that sort of thing." (DoD 17)
	"Small business is another one which is a subset of the industry perspectives. And in fact if you look at—you may want to make note of that one because as part of this discussion, sometimes when we're talking about business acumen and industry [knowledge gaps], we risk overlooking the small-business piece." (DoD 8)

Table 3.2—continued

Type of Knowledge Gap	Exemplary Quotes
Agile development	"The agile development I think, again, industry has long moved away from waterfall planning. . . . There's this interactive and agile approach. I think DoD is aware of it but I don't think it's, again, permeated through the entire ecosystem, if you will." (Industry 11)
	"I taught computer programming when it was Fortran. So I don't know what Tactical Edge Computing is but she [IT SME] is identifying those areas like that to say, 'This is critical. We need to bring industry in and help teach our people how they do things.' Agile development, same thing, she mentioned that." (DoD 2)
Cybersecurity	"And then another hot topic that we think there's a gap that's increasingly important is cyber. . . . I think it's under [industry] operations. That may be part of the reason that we're going there is there's been significant development of cyber learning assets here but, you know, a lot of that has been recent because [it's] a kind of recent focus." (DoD 1)
	"Cybersecurity has been on our list for a while. Defense Business Systems has been on our list for a couple of years. Agile software development, you know, getting systems out sooner, is something that—you mentioned gaps—the Defense Science Board issued reports saying that we're not getting that out fast enough." (DoD 22)
Rapid pace	"The corporate world is changing so quickly that I think DoD is really struggling with how do they keep up. And if you're going to listen to any DoD leader, there's a lot of talk about acquisition reform and the days of taking decades to build a tank, you can't buy technical solutions the same way you bought industrial solutions. So having that perspective of how quickly industry operates and how quickly we iterate and how we run engagements and run projects I think is tremendously valuable for the Acquisition Corps to see, have that, gain that perspective because they probably aren't seeing it." (Industry 11)
	"We need to constantly be at a point of revolutionizing what we do, what we learn, the way we do things. If you think about you go from a whole bunch of CDs to now nobody has CDs. . . . But that's the way it is in a lot of [career field] functions that things are passing us by and that's because I think there is a tendency for us not to accept technology as much as we could. And we're doing a lot in engineering in terms of using three-dimensional modeling, using 3D printers, but there's a lot more we could do, I think. So that to me is just a challenge within the career field and I remember that's something . . . you get from the experience and exposure to different things." (DoD 14)
	"There are countless gaps, particularly in the industry operations piece. Not just technology in particular areas, it's just moving so fast. I tell you, it's tough to keep up. It is tough to keep up across all these disciplines. And we haven't even started talking a lot about IT yet . . . with what's going on out there. And we're not refreshing our workforce. . . . The average age I think in the engineering workforce is about 46 years old. And they're not being refreshed with any frequency. So there are countless gaps in the industry operations area." (DoD 10)

SOURCE: 2018 RAND Section 843 study interviews.

of these gaps. The strongest evidence in terms of frequency, richness of discussion, and agreement across interviewees was related to industry financial practices—the nature of their financial management operations, how to interpret financial documents, an understanding of industry accounting, and the like. Supply chain management was another area seen as an aspect of business operations of which AWF needed greater awareness. Knowledge of small business operations was perceived as important because small businesses typically have limited resources to meet government financial and legal requirements. Yet this knowledge was potentially lacking, possibly due in part to a focus on larger-scale business operations.

The next two aspects of industry operations, agile development and cybersecurity, pertained to techno-business knowledge. According to DAU, agile development is "a set of methods and practices based upon the values and principles of the Agile Manifesto. Through self-organizing, cross functional teams, software is rapidly and iteratively developed in response to evolving requirements."[17] Cybersecurity is "[p]revention of damage to, protection of, and restoration of computers, electronic communications systems, electronic communications services, wire communication, and electronic communication, including information contained therein, to ensure its availability, integrity, authentication, confidentiality, and nonrepudiation."[18] Although both topics are sometimes regarded as aspects of technical management or technical competencies rather than business-related aspects, we include them here because interviewees stressed emulating how industry operates or following its lead. For example, one DoD leader stated:

> I would consider industry operations to include common technical processes that they use in the execution of their work. The greatest example I can think of in industry operations that we can wrap our heads around is the idea of modern software development methodology. So I would consider the increasing use of agile software development . . . . [T]hat could be an example of a current industry operation. (DoD 10)

Finally, interviewees referred to a gap that cut across industry operations: an awareness of and appreciation for the fast rate of change in industry and the resultant ability of leading companies to adapt.

In our search for corroborating evidence, we found that knowledge of industry operations was assessed in two DoD studies, a study about the Program Management career field with publicly available results and one unpublished, service-specific DoD study. The small number might be because the studies we reviewed mainly pertained to skills and competencies and did not call out specific knowledge types on their own or as part of competencies. This could be a problem because, as one DoD leader told

---

[17]   DAU, *Agile Software Development*, undated(b).

[18]   DoD, *Cybersecurity*, DoDI 8500.01, March 14, 2014, p. 55.

us, "I would argue this [Section 843–related knowledge] is part of the domain knowledge of understanding your industry partners, not a competency" (DoD 16). The lack of studies also may be due to the tendency to focus on technical competencies in AWF career-field competency models rather than "soft" competencies. In the DoD 2014 *Study of Program Manager Training and Experience*, most program managers surveyed (51 percent) either felt that acquisition training was insufficient or were uncertain about its ability to help them understand and use contractor financial reports. Although it is possible that they could have acquired this knowledge in a different way to compensate for the training they believed was inadequate, there may have been a knowledge gap in this area at the time of the study.[19] The unpublished, service-specific DoD study identified gaps present in one career field in relevant competencies, such as

- the differences in business operations and strategies between companies focused on weapon system commercial products and services
- business functions in a company's organization, such as marketing, finance, operations, human resources, and accounting
- how company elements vary in importance over a program life cycle and the key elements of each phase
- how companies break down and utilize overhead pool structures
- the interrelationship between significant government and company decisions, a company's business strategy, and financial capability assessments
- the impact of supplier management on company margins and program performance
- company management of critical supply chain priorities within and across programs.

In some cases, the lack of proficiency in these and other, similar competencies was only documented at the junior level, but some, such as the interrelationship between significant government and company decisions, were found at junior, mid, and senior career levels. However, that study was about five years old at the time of this report's publication, and it is possible that DoD efforts to close some of these gaps through DAU courses and other means have been at least partially successful.

We also noted some studies that were not expressly focused on knowledge gaps but offer some support for the need for greater knowledge of industry operations related to cybersecurity and supply chain management. For example, cybersecurity was an area of focus in GAO's 2017 high-risk area report, with cybersecurity seen as a mission-critical skill gap cutting across federal agencies. Some of this stemmed from an insufficient number of cybersecurity-focused professionals versus a knowledge gap related to how industry addresses cybersecurity. In the same report, GAO also cited DoD supply

---

[19] O'Donnell, 2018.

chain management as a high-risk area, albeit one in which improvements were noted. Again, although knowledge gaps were not expressly cited, personnel issues were seen as part of the problem—and part of the solutions being implemented.[20]

Finally, we heard in some interviews that gaps in business acumen, most notably risk management and EVM, may be amplified by a lack of knowledge of industry operations related to these specific activities. As one interviewee told us:

> DoD personnel may learn what Earned Value Management is, but they don't understand what leadership in the best organizations actually do to apply Earned Value Management within their organizations. And DoD approaches to project management don't necessarily reflect what industry views as the best approaches to project management. (Industry 4)

Essentially, awareness and understanding of the actual practices industry uses in context was regarded as critical to successful application of the business acumen "basics."

Gaps Related to Knowledge of Industry Motivation

> I think their [industry's] motivation is obviously a little bit different than you would find within the Department of Defense, and we consistently talk about those motivations as being important to your understanding for folks in the acquisition business. Because learning what motivates industry is important to how you structure and create a win-win deal or a contract with that company. So we find that all too often, and in my previous experience, is that you'll probably find that clarity and understanding relative to industry motivation is not clear and I would suggest is even lacking in general inside of the Defense Department and probably inside of the federal government. And that is, indeed, a challenge. (Industry 7)

Similar to industry operations–related knowledge, our interviews included rich discussions of both the importance of and the greater need for knowledge of industry motivation within the AWF. However, it was covered in only one of the studies we reviewed, which was the unpublished study carried out by one service in which one career field's competency gaps were identified. The same possible reasons for fewer assessments of gaps in industry operations knowledge seem relevant here as well. In that study, personnel lacked proficiency in industry motivation–related competencies, such as

- incentives that drive desired decisions and behaviors of the company

---

[20] GAO, 2017.

- the elements that drive compensation for chief executive officers, project managers, business developers, and capture managers (for new business)
- key corporate motivators that ensure meeting financial objectives and reliable cash flow, profit, and growth opportunities
- government actions that affect company planned financial forecasted revenues
- industry motivations and advantages from using different types of cost estimates
- factors driving industry management team pressures to "make their numbers"
- the interests of key company stakeholders, including shareholders, debt holders, boards of directors, executive management, and business units
- the interests of company program managers, control account managers, operations, and engineering teams.

As noted in the industry operations discussed, the results of this study were broken out by junior-, mid-, and senior-level personnel. Perhaps most significantly, "incentives that drive desired decisions and behaviors of the company" was a knowledge gap for personnel at all career stages.

Although this study is a bit dated, interviewees' comments about understanding industry motives and incentives align with many of those competencies. For example, the sentiments that follow describe AWF challenges in understanding the incentives that influence industry decisions and actions:

> If we had a better understanding of industry motivations, we might actually write better incentives to get the products that we're looking for under the conditions in which we want them. I suspect that one of our challenges is . . . we write the incentives that we think are important. We don't necessarily write the incentives that industry thinks are important, that we can also leverage to get better performance. (DoD 20)

> The two primary goals that most organizations have [are] staying financially viable and providing good products. And it's not that one is necessarily more important than the other, but profit and all of the things an organization does to stay financially healthy are really important to driving the way they do business. And people coming up in government don't understand anything about this. They don't understand that if you build something as a fixed price contract, someone has to account for the risk, so you will get a much higher bid than if you made it cost plus fee. (Industry 4)

> So much of that [understanding industry] means understanding the nature of the risk that our work brings to a vendor and what the vendor can take on and what we can take on. . . . [I]f we misjudge the risk, we end up with readiness shortfalls. If the vendor misjudges it or we convey it in such a way that it's confusing, we usually end up with both unintended cost increases or risk or both. So that acumen in terms of . . . not just structuring the contract, [but also] once the contract is in

place, the vendor is going to respond to whatever you put in the contract in the way of feedback mechanisms and how the contract was structured in terms of incentives and fees, fixed price, cost-reimbursable, the time, the period of performance associated with it. All of those things I think play into obviously how the contractor is going to respond but we need to understand it I think better throughout our workforce. (DoD 15)

The second and third comment above also relate to the "industry motivations and advantages from using different types of cost estimates" competency listed above. Another theme in the interviews consistent with the study findings pertains to the compensation of CEOs and other corporate leaders, as the following comments illustrate:

A lot of times what motivates an industry is how the CEO is compensated. So, I would tell you we probably don't do a good enough job, in my opinion, of focusing on that. (DoD 16)

There's a huge difference between compensation in government and industry. And it's not a difficult concept. We don't have to spend a lot of time on this, but we do need to explain that the guy sitting across the table from you, he's not paid on the rank. He's paid on results, and these are the results that matter and this is the way their compensation package is structured and you need to understand the consequences of that, not just for the top organization but all the way down to the program management office, how are they being evaluated and compensated and incentivized. (Industry 15)

These findings suggest that, at least according to some of the SMEs we interviewed, the gaps in knowledge of industry motivation that one of the military services assessed five years ago may still pose a problem for at least part of the AWF.

Ripple Effect of Industry Knowledge Gaps

While our study was focused on the three specific types of business knowledge specified in Section 843, interviewees mentioned other types of business-related knowledge gaps that could be influenced by knowledge of industry operations and industry motivation—or a lack thereof. **Negotiation** is a compelling example of this type of issue. It was cited as a gap in several studies, including three of the unpublished DoD studies we reviewed and the 2018 Professional Services Council (PSC) Acquisition Policy Survey. In the 2018 PSC survey, not only did results indicate that the federal acquisition workforce (including but not limited to DoD) found negotiation to be challenging, but 75 percent of the 65 survey participants also reported that negotiation skill levels had not changed in the preceding two years. Further, 43 percent of participants felt that this situation would not change in the next two to three years.[21] Our interviewees mentioned negotiation in their discussions of Section 843 knowledge–related gaps and offered observations such as the following to explain how a lack of industry-related knowledge contributed to this problem:

> The Department teaches the hard skill: Here's what the FAR provides, here's the contract type, here's how you build a contract file. They rarely talk about elements of negotiation or understanding of commercial business operations or activities, and that's where we continue to see that skills gap. And it's a significant one because . . . if you don't understand who you're negotiating with, then all you can fall back on is what you've been told or what you know for your own, and that hamstrings the Department's approach. (Industry 2)

> Frankly, I think one of the big things for folks who have never gone through any extensive exposure to industry or at least dialogue or training with respect to this topic don't really get why they [commercial firms] do what they do and, because of that, don't know how to sit properly across the negotiating table to find common ground, right? I mean, if you realize that Company A [wants to meet] their quarterly projections and that if you can get a deal done before the quarter is over, you can get a much better deal. So many people don't get that, right? But that's a huge thing. (DoD 5)

> Industry is more about how do they make profit, right? And so, understanding their cost structures and how direct and indirect labor works and all that stuff is really important when you're negotiating a contract. (DoD 16)

Developing and understanding requirements emerged as another knowledge gap that could be improved or that exacerbated an individual's level of industry-related knowledge. Developing requirements and scopes of work for services was identified in the 2018 PSC survey as an ability of critical importance, but survey respondents tended to perceive federal acquisition workforce skills levels as fair in this area. In a related vein, only about half of program managers who participated in DoD's 2014 *Study of Program Manager Training and Experience* reported that the acquisition training they received to respond to user requirements was sufficient, and proficiency in this area was reported as needing improvement in a recent unpublished DoD study that we reviewed. Again, our interviews provide insight on how knowledge of industry operations and motivation plays a role:

> People who are writing requirements are not those people who are very well informed at all on what the technological state of play is. So they write require-

---

[21] PSC, *Optimism Amid Diversity: The 9th Biennial Professional Services Council Acquisition Policy Survey*, July 2018.

ments that are very ambitious. They go through a government vetting process which is basically run by other people who are not intimately familiar with the state of technology and then you get a requirements document that comes out that is something that is too ambitious for industry to actually provide in the cost and on the schedule that the government wants. (DoD 6)

A third gap potentially influenced by knowledge of industry was related to **cost and price analysis**. This area was another challenge for the federal acquisition workforce cited in the 2018 PSC survey,[22] and exemplary quotes from the interviews include the following:

> The world of software is probably the best example. It's moving so fast that it's very hard for us to keep up with the latest and greatest in cloud computing and how do you price stuff like that . . . those are hard things to keep up with. (DoD 23)

> I would say . . . how contractors price their proposals [is a gap]. I mean, wouldn't that be wonderful to know how they do it, what kind of risk gaps they put in there; is it 10 percent, is it 20 percent? You know, what other allowances did they include in their proposals? How did they determine risk to a project? (DoD 24)

There may be other types of knowledge affected by the level of business acumen, knowledge of industry operations, and knowledge of industry motivation that an acquisition professional possesses. The ones presented here were salient to the interviewees and consistent with evidence from other studies. If future competency or knowledge proficiency assessments include the concepts enumerated in Section 843, that would enable DoD to understand better how much these "soft" knowledge domains may come to bear about successful applications of other business-related knowledge.

Summary

Our interviews revealed a great deal of variation in the timing and methods DoD uses to assess AWF proficiency in different types of competencies. This may change with the new initiative to conduct career field–level assessments using DCAT, but it is unclear whether business acumen, knowledge of industry operations, and knowledge of industry motivation will be included in the DCAT approach. Nor is apparent how the DCAT approach and results relate to FIPT efforts to build, revise, and maintain career field–level competency models.

The dearth of knowledge gaps assessments at the time of this study, coupled with the absence of standards related to business acumen, knowledge of industry operations, and knowledge of industry motivation, rendered difficult a systematic, detailed

---

[22] PSC, 2018.

examination of the knowledge gaps focused on in Section 843. Through a combination of expert interviews and review of a small set of recent, relevant studies, we determined that knowledge gaps related to business acumen exist within the AWF, particularly those related to risk management and EVM. We also identified gaps in knowledge of industry operations, including financial aspects, supply chain management, small business, agile development, and cybersecurity. Industry's rapid rate of action and adaptation was also cited as a type of knowledge of industry operations that the AWF did not sufficiently possess, and interviewees noted that a lack of awareness of how industry uses EVM and conducts risk management limited acquisition professionals' application of those forms of business acumen, With respect to industry motivation, interviewees felt that this was an important type of knowledge lacking within the AWF workforce, including aspects such as the incentives that drive desired decisions and behaviors of commercial firms as well as the influence of corporate executive compensation. Insufficient knowledge of industry was also seen as contributing to gaps related to negotiation, developing and understanding requirements, and cost and price analysis.

Our review of competency models suggests that some of the knowledge type gaps we covered in this chapter, such as those related to risk management and negotiation, are perceived as needed by most career fields. But the analysis limitations noted earlier prevented us from specifying which career fields suffered from these gaps and to what extent. In addition, some of the studies we used in our analysis were completed in the 2013–2017 time frame (i.e., not immediately before this study), which means that DoD has already taken actions to close these knowledge gaps and may have made progress in this regard. We highlight some of DoD's efforts to close specific gaps in the next chapter.

CHAPTER FOUR

The Use of Training and Development to Address Knowledge Requirements

The study parameters specified in Section 843 included an assessment of DoD's current use of T&D to instill business acumen, knowledge of industry operations, and knowledge of industry motivation in the AWF, with special attention to whether external T&D resources might be better utilized to address knowledge gaps in these areas. To determine whether more external T&D should be utilized, there are many types of information needed as criteria, including the range of T&D resources available, the current use of internal and external T&D resources, the costs and benefits of various internal and external T&D options, and the barriers and facilitators to the use of external T&D. This chapter features a high-level compilation of those criteria. We start with a description of the various types and providers of T&D that are available to DoD, followed by a description of the primary sources of internal and external T&D that DoD is currently using to build knowledge of business acumen, industry operations, and industry motivation. We then summarize findings on the benefits that external T&D resources offer to AWF personnel, followed by a discussion of the facilitators of, and barriers to, the use of external T&D resources. Finally, we conclude with a discussion of whether there is a need for additional use of external T&D.

Options for T&D to Build Business Acumen, Knowledge of Industry Operations, and Knowledge of Industry Motivation

To set the stage for our discussion of DoD's portfolio of internal and external T&D offerings, we conducted a scan of the T&D environment to describe the different types of T&D that DoD might draw on to address knowledge requirements related to business acumen, industry operations, and industry motivation. The type of T&D is an important first decision that must be considered before assessing the landscape of providers and determining whether external resources will be used. The optimal form of T&D for addressing knowledge requirements may vary by type of knowledge, and various forms of T&D may need to be used together to ensure that knowledge translates into the desired changes in behaviors and organizational outcomes.

We focused on a review of the executive T&D literature and our interviews as the sources for describing the landscape of T&D options available and settled on six:

- certifications
- training courses
- executive education programs
- degree programs
- on-the-job training (OJT)
- rotational assignments.

We also examined descriptions of T&D for a set of eight large corporations: Amazon, Boeing, Deloitte, Ernst & Young, Google, Lockheed Martin, Microsoft, and Raytheon.

Certifications

Certifications and certification-related T&D play an important role in developing knowledge for many career fields. Certifications can be divided into two groups: professional certifications and occupational certifications. In many career fields—not just those related to acquisition—professional associations have developed certifications that help individuals within a profession to develop a core body of knowledge and to signal the mastery of this knowledge to employers. Professional certifications typically require individuals to take a test that assesses a core body of knowledge. Associations often offer courses and training resources to support individuals in preparing for the certification assessment and require individuals to engage in continuing education and training efforts to maintain the certification. Some professional certifications are held by many individuals within a certain career field. For example, the Project Management Professional (PMP) certification offered by the PMI is held by approximately 750,000 project and program managers worldwide. The National Contract Management Association (NCMA) is an organization that offers certifications for contracting professionals, such as Certified Professional Contract Manager (CPCM), which is held by approximately 3,000 individuals in the contracting field.[1] Some career fields do not have an association or certification that is perceived as setting the industry standard for the field. These career fields may have smaller niche associations and certifications held by a small proportion of individuals in the field.

In addition to professional certifications offered by professional associations, many education and training providers, such as colleges, universities, and training centers, offer occupational certificates to signal that individuals have completed a series of courses preparing them for a particular field. DAWIA certifications fall into this

---

[1]   From an October 19, 2018, phone call with the Manager of Certification Operations at NCMA. NCMA also offers the Certified Federal Contract Manager (CFCM) certification, which is also held by about 3,000 people, and the Certified Commercial Contract Manager (CCCM) certification, which is held by about 450 people.

category. A DAWIA certification level does not expire, but individuals are required to accomplish continuing education in order to maintain currency.[2]

Training Courses

Training courses, including classroom-based and online courses, are another form of T&D that could be used to address requirements for Section 843–related knowledge. A wide range of providers offer training courses that might help to build business acumen and knowledge of industry, including two-year and four-year educational institutions, commercial vendors, and corporate universities at private-sector organizations. Some providers offer only prepackaged courses, while others offer opportunities to modify and tailor training content for particular needs of the organization. Private-sector organizations are also increasingly drawing on open-source T&D resources.[3]

According to a 2016 study by the Association for Training and Development, "about half of organizations still concentrate design efforts on traditional classroom instruction and e-learning."[4] However, the use of traditional training courses is somewhat less common for the purposes of training higher-level executives. A 2014 survey of executive T&D by Executive Development Associates (EDA) found that internally and externally designed online training programs and off-the-shelf training were the least commonly used approaches, used by just 2 to 8 percent of organizations surveyed, depending on type.[5]

Executive Education Programs

According to a 2017 study of executive education programs, the goals of these programs are "improvement in the skills, knowledge, and abilities needed to become an effective leader."[6] These programs typically last anywhere from one to six days, with multiday programs often requiring residency on campus.[7] Executive education programs are designed to be intensive and immersive experiences that mix a variety of different learning approaches, with an emphasis on experiential learning.[8] Some programs are offered in a modular format where participants are expected to apply T&D on the job

---

[2]   Personnel at every DAWIA certification level must earn 80 hours of continuous learning credit every two years (DoDI 5000.66).

[3]   S. Herring, "MOOCs Come of Age," *T+D*, Vol. 68, No. 1, 2014, pp. 46–49.

[4]   Association for Talent Development, *Experiential Learning for Leaders: Action Learning, On-the-Job Learning, Serious Games, and Simulations*, Alexandria, Va., 2016b.

[5]   Executive Development Associates, undated.

[6]   W. W. Stanton and A. D. Stanton, "Traditional and Online Learning in Executive Education: How Both Will Survive and Thrive," *Decision Sciences Journal of Innovative Education*, Vol. 15, No. 1, 2017, pp. 8–24.

[7]   Stanton and Stanton, 2017.

[8]   Executive Development Associates, undated.

in between more-intensive days of on-site T&D.[9] A 2014 survey found that 19 percent of organizations identified university-offered executive education programs as a top form of T&D used for corporate-level ("C-suite") executives, and 8 percent of organizations used them as a top form of T&D for high-potential managers.[10] While business schools are a primary provider of executive education programs, they are also offered in house by corporate universities and by other organizations, such as the Aspen Institute.

Many business schools and commercial vendors also offer tailored courses and executive programs for companies that are looking to address a more-customized business need that cannot be addressed by existing programs and courses.[11] Approximately 8 percent of organizations in the 2014 EDA survey reported using custom-designed courses from universities as a main form of T&D for top executives.[12]

Degree Programs

For a deeper set of knowledge, skills, and abilities on business-related competencies, organizations can hire personnel who already have degrees, or they can provide individuals with opportunities to pursue undergraduate and graduate degree programs while working for the organization. According to a study by Bersin of Deloitte, 71 percent of organizations offer programs that assist with tuition.[13] These tuition programs could be used for deep training on particular business fields (e.g., cost estimating, supply chain logistics) or more general programs for executives, like the Executive MBA degree. Executive MBAs typically require part-time participation over one to two years and have restricted admission.[14] In addition to participation in standardized degree programs that are offered to individuals across fields and organizations, larger organizations may also work with business schools to develop tailored degree programs for individuals in specialized fields like acquisitions.

On-the-Job Training

A 2010 McKinsey survey found that the most common training method used "extensively" is OJT, and more-formal types of training are somewhat less common.[15] The executive T&D literature also heavily focuses on on-the-job approaches to T&D as being important and among those most commonly used, with OJT including 360-

---

[9]   Stanton and Stanton, 2017.

[10]   Executive Development Associates, undated.

[11]   S. Perez, *The ROI of Talent Development*, Chapel Hill, N.C.: University of North Carolina, Keenan-Flagler Business School, 2014.

[12]   Executive Development Associates, undated.

[13]   Bersin, *Tuition Assistance Programs: Best Practices for Maximizing a Key Talent Investment*, Oakland, Calif.: Bersin by Deloitte, 2012.

[14]   Stanton and Stanton, 2017.

[15]   McKinsey, *Building Organizational Capabilities: McKinsey Global Survey Results*, 2010.

degree feedback, coaching, and mentoring.[16] In the EDA survey mentioned above, executive coaching was the most commonly used form of T&D for C-suite executives. Executive coaching was reported as a primary T&D strategy by 54 percent of organizations, and 29 percent of organizations reported using executive coaching for high-potential managers.[17] Mentoring was reported as a primary strategy by 26 percent of organizations for developing C-suite executives and was reported as a primary strategy by 34 percent of organizations for developing high-potential managers.[18]

Rotational Assignments

Rotations are another form of T&D often used by organizations to develop personnel. Rotations can last anywhere from a few weeks to a year or more and are intended to expose participants to practices in other departments (when conducted internally) or to the practices of other companies (if conducted externally), as well as to foster cross-organization collaboration and understanding more generally. There was no discussion in the literature on external rotation programs, suggesting that they are much less common than internal rotations. A 2014 survey found that 19 percent of organizations reported "developmental job assignments" to be a primary form of T&D for C-suite executives, and 46 percent reported the job assignments to be a top form of T&D for high-potential managers.[19] Another study found that 12 percent of organizations offered their high-potential employees job rotations, while 25 percent offered stretch/special assignments.[20]

DoD's Current T&D Portfolio Addressing Business Acumen, Knowledge of Industry Operations, and Industry Motivation

To address knowledge requirements for business acumen, industry operations, and industry motivation, DoD currently employs a range of different T&D opportunities. We first describe the internal opportunities for T&D that are intended to confer these specific areas of knowledge, followed by a description of external T&D opportunities available to the AWF and the resources that help personnel to access these external opportunities.

---

[16] Executive Development Associates, undated.

[17] Executive Development Associates, undated.

[18] Executive Development Associates, undated.

[19] Executive Development Associates, undated.

[20] J. Filipkowski, *Accelerating Leadership Development*, Chapel Hill, N.C.: Human Capital Institute and UNC Kenan-Flagler Business School, 2014.

Internal T&D Options Currently in Use by the AWF

DoD relies heavily on internal T&D options to ensure that the knowledge requirements of the acquisition workforce are met. Table 4.1 shows one way of looking at three major categories of DoD providers of this training: DAU, other DoD and service schools, and training offered by individual services. We will consider each of these categories in turn and highlight the modes of T&D that address business acumen, knowledge of industry operations, and knowledge of industry motivation.

Defense Acquisition University

The DAU president is the Chief Learning Officer for the acquisition workforce[21] and directs all education activities provided by the university. DAU was established to provide for the professional educational development and training of the AWF and delivers training courses for each of the acquisition career fields that allow AWF members to be DAWIA certified at Level I, Level II, or Level III.[22] In Chapter Two, we identified several courses required for certification in one or more career fields that address the areas of knowledge highlighted in Section 843, among them the "Understanding Industry" course, ACQ 315. According to a 2015 GAO report,[23] the course was developed in response to two studies conducted by the Program Management career field to identify opportunities to improve the proficiency of program managers through additional training. ACQ 315 is required for Level III certification in Program Management and is one of several courses that meet a Level III requirement for two other career fields. The course addresses a wide range of topics related to the types of knowledge that Section 843 focuses on, including business strategy and development, incentives, negotiating strategies, and operations that motivate company decisions to meet their business goals. Table 4.2 shows that the course had 1,143 civilian and 452 military graduates from all AWF career fields in FY 2018, except for Property Management.

Table 4.1
Overview of DoD T&D Options

Internal Provider of T&D	Certification	Courses	Executive Education	Degree	OJT/ Rotations
DAU	X	X	X		
Service schools		X		X	
Services		X			X

---

[21] DoD, DoDI 5000.66, par. 3.2, 2017a.

[22] DAU, undated(g), pp. 14, 36.

[23] The studies, conducted in 2009 and 2014, "identified, among other things, the need to improve program managers' awareness of earned value management . . . and business acumen" (GAO, 2015, p. 19).

Table 4.2
ACQ 315 Graduates in FY 2018

Career Field	FY 2018 Graduates	
	Civilian	Military
Auditing	1	0
Business-Cost Estimating	1	0
Business-Financial Management	12	0
Contracting	193	36
Engineering	53	11
Facilities Engineering	1	0
Information Technology	8	1
Life Cycle Logistics	148	18
PQM	6	7
Program Management	613	322
Purchasing	1	0
Science and Technology Management	1	1
Test and Evaluation	4	8
Unknown[a]	101	48
Total	1,143	452

SOURCE: ACQ 315 student graduation data provided by DAU.

[a] ACQ 315 allows "walk-in" students to fill last-minute vacancies in the course. These students do not register through the standard process, and so career-field information about them is not always available.

In addition to courses required for certification, DAU has also developed courses to address perceived knowledge deficiencies present in the AWF, including those related to the knowledge gaps described in Chapter Three. For example, in response to congressional direction, DAU created classes to address perceived needs for more training on agile software development. One of our interviewees described the process as follows:

They [Congress] specifically wrote a requirement in the FY18 NDAA for us [DoD] to develop a classroom course on agile software development. So poof and 90 days later, we got a course and we're using, we're launching very heavily industry source material. Actually, to stand it up very quickly, DAU licensed some industry source material that's used in training students in the commercial world. And so DAU can make that happen very quickly, if they make it a requirement in the National Defense Authorization Act or if Miss Lord or whoever of the functional leaders make it a requirement for their career fields. (DoD 21)

Another new class that may help to address the cost and pricing analysis gap described in Chapter Three is "Applied Software Cost Estimating," BCF 250. This class was initially developed in response to a training gap identified in the 2014–2015 time frame and was piloted in FY 2017. As of FY 2018, all personnel in the Business-Cost Estimating career field seeking Level II certification were required to successfully complete this course.[24]

In a proactive approach to avoid gaps in Section 843–related knowledge and to develop acquisition professionals more generally, DAU also offers in-residence and online "core plus" courses (topics that are considered important, but not required, for a certification level) and continuous learning options that help AWF personnel satisfy the requirement to complete 80 continuous learning points every two years.[25] As an example of how these courses address the areas of interest cited in Section 843, DAU has secured access to a series of Harvard Business Publishing product suite educational modules, such as "Finance Essentials" (HBS 417), which introduces non-financial managers to income statements, balance sheets, and cash-flow statements from an industry perspective. These modules are part of DAU's Knowledge Repository, which is accessible to the entire AWF.[26]

According to our interviews, an effort is made to incorporate an industry perspective into appropriate courses:

> We're constantly interfacing with industry trade groups like NDIA [National Defense Industrial Association], AIA [Aerospace Industries Association], PMI. We're on the Project Management Institute Global Executive Council. We're actually members of that Global Executive Council, DAU. . . . So we're parts of all these different industry groups and we go to the meetings, we try to network and leverage that information as much as possible. As I said, we get our own certifications outside of DAU. It's not so much—it really just didn't expand my knowledge any but it helped give me insight into how industry is training their people and where maybe we could do something different." (DoD 21)

---

[24] R. P. Burke and N. Spruill, *Implementation Memo to Add a Core Certification Course for the Business—Cost Estimating Career Field*, memorandum from Richard P. Burke, Deputy Director, Cost Assessment and Nancy Spruill. Director Acquisition Resources and Analysis, April 15, 2016. Although this course is intended to address a gap that is indirectly related to Section 843 knowledge, we did not include it in the analysis described in Chapter Two because the course objectives do not refer to industry operations.

[25] Courses that are required for one career field may be "core plus" courses for another career field or may be neither required nor "core plus" courses but nonetheless count toward meeting continuous learning requirements.

[26] See DAU, undated(g), p. 8, for Knowledge Repository and p. 207 for HBS 417.

In addition, professionals from industry—181 in FY 2017[27]—participate in DAU courses as students, which can help give government participants perspective on business acumen, industry operations, and industry motivation:

> But probably the one gap is that the DAU stuff does tend to focus more on the government side, as opposed to kind of understanding the whole system. So that's why they always like to have people in classrooms that are from industry because then you can talk about both sides. You can say, "Oh this was the government side of the proposal," and the contractor says, "When you idiots made us work Thanksgiving and Christmas because you didn't even think about what it was like to be a contractor." (DoD 18)

Efforts are also made to include guest speakers who provide a non-DoD perspective:

> [W]e intentionally bring in speakers who are not necessarily internal to Army or DoD, right? Because we recognize that a broader perspective on how to solve problems to include industries that you wouldn't normally think of. . . . if you look at problems through a different lens that they look at it through, you could be more creative on how you solve them." (DoD 5)

> You know, many of the courses that DAU offers include industry speakers that give you the perspective of, you know, what it's like to run a small business and, you know, what it's like to be on the other side. Now, hearing it with a speech or a PowerPoint chart is not the same as seeing it and living it, but there are those, you know, lighter touches that exist. (Industry 18)

DAU provides what could be termed executive training: short, intensive courses for more-senior leaders. One example is the "Executive Refresher Course" (ACQ 405), the goal of which is to update senior acquisition professionals on acquisition policies, processes, and lessons learned and includes industry guest speakers.[28] Another is ACQ 415, "Strategic Interface with Industry." This three-day in-residence course is meant to provide focused, comprehensive business knowledge training for senior acquisition personnel (SES; senior GS-15 or military O-6 or above).[29]

---

[27] *RAND Industry Grad Request FY17* spreadsheet provided by DAU, August 2018. The figure reported herein is the number of industry professionals who participated in classroom-based education offered by DAU. Additional industry professionals availed themselves of DAU's web-based learning assets, but we did not include those numbers, reasoning that web-based education does not afford much opportunity for industry-government interaction.

[28] DAU, undated(g), p. 135.

[29] DAU, undated(g), p. 135. One of our interviewees (DoD 21) noted that this was an example of DAU recognizing a need for a course that was not recognized through the competency model process.

DAU also provides what it calls "performance learning," which takes training to the workplace via mission assistance events and workshops. Mission assistance events include short-term and long-term consulting engagements, Major Defense Acquisition Program/Major Automated Information System support, collaborative problem-solving events, and rotational assignments to program offices.[30] An example of this training that addresses the areas of knowledge of interest to Congress is the "Understanding Industry Workshop" (WSM 016), which is designed for people who are unable to attend ACQ 315 in residence. Mission assistance programs have been used to address specific, recognized gaps in knowledge:

> In terms of Cyber Security, DAU is at this point taking multiple approaches and we've been supporting that, in terms of the DAU Mission Assistance Program and workshops that are in the Cyber Security area. So there are multiple ways to reach the Acquisition Workforce and the certification courses is one major venue and then the distance learning options is another. But more and more, DAU is looking to make direct contact with programs, work with programs through Mission Assistance, as well as to establish workshops that are topic-based, that are a direct use to programs. (DoD 22)

Other DoD Service Schools

In addition to DAU, DoD provides graduate education opportunities and other training that address business acumen, industry operations, and industry motivation through other DoD-operated schools. The schools emphasized on DACM websites and in our interviews were the Naval Postgraduate School (NPS), the Air Force Institute of Technology (AFIT), and the Eisenhower School for National Security and Research Strategy. Both NPS and AFIT award degrees in areas related to industry operations, such as the Master of Science in Contract Management (MSCM) at NPS and the Master of Science in Cost Analysis at AFIT.

Opportunities to participate in these programs are relatively limited, as demonstrated by the fact that the NPS MSCM degree is a distance-learning course in the Graduate School of Business and Public Policy (GSBPP) program,[31] and the total number of distance-learning graduates in the GSBPP program overall in 2015 was

---

[30]   DAU provided 403 mission assistance events in FY 2017 that provided 172,013 "contact hours" with participants. (See DAU, *Shaping the Future: 2017 Annual Report*, undated[g], p. 27.)

[31]   We highlight the NPS Master of Science in Contract Management degree because it was specifically mentioned in our interviews. The GSBPP offers two resident programs: an MBA program and a Master of Science in Management. It also offers two other distance learning programs: a Master of Science in Program Management and an Executive MBA. These programs are described at the NPS website (NPS, "Degree Programs," undated[b]). According to the website, the Master of Science in Contract Management satisfies requirements for DAWIA level III in the contracting career field, and the Master of Science in Program Management fulfills several DAWIA requirements.

106.[32] AFIT had only three Master of Science in Cost Analysis graduates in 2015 (the latest information available on its website).[33] AFIT also provides individual continuation training courses, such as "Industry Standard Project Management,"[34] which educates students on the differences between Air Force and industry project management.

The Dwight D. Eisenhower School for National Security and Resource Strategy (formerly the Industrial College of the Armed Forces) under the National Defense University (NDU) graduates over 300 students a year with Master's degrees in national resource strategy. Graduates tend to be senior military and civilian personnel (with 20 years or so of service), but the program also includes a small number of fellows from industry.[35] A select number of personnel attend the school as part of DAU's "Senior Acquisition Course" (ACQ 401), which is designed to prepare military officers and civilians for senior leadership and staff positions throughout the acquisition community.[36]

Training Offered by Services

The individual services provide acquisition training to address needs that may be unique to their organizations. This training includes courses designed to meet service-specific needs, OJT, and internal rotations (exposure to different jobs within the service).

Naval Air Systems Command, for example, created what it calls NAVAIR University in 2013.[37] Online training through eight "colleges" is available to virtually anyone in the NAVAIR community,[38] and, according to one of our interviewees, it helps address the topics raised in Section 843:

> [In] the NAVAIR University career guide they offer courses in working effectively with industry. (DoD 2)

---

[32] Figures are not available by degree type. Details for 2015 are available at NPS, "Dept. Graduation Rates: GSBPP (GB)—All Students," April 8, 2016. In 2017, there were 440 total NPS distance learning graduates. See graduation statistics at NPS, "NPS Degrees Conferred by Academic Year, Quarter and Type of Enrollment: Graduation AY 2000 to 2017," May 2018.

[33] AFIT, "AFIT Graduates Class of 248, March 2014," April 2, 2014.

[34] See course description on the AFIT website at AFIT, "Course Information and Registration," undated.

[35] The Eisenhower School, "Students," undated(b).

[36] See the Eisenhower School, "Departments," undated(a).

[37] At least as late as 2012, the Air Force's Space and Missile Systems Center (SMC) had an "SMC University" that provided training in Section 843–related areas, such as contracting, financial management, and program management (Gallagher, 2012).

[38] Colleges of Business Financial Management and Comptroller, Program Management, Contracts Management, Information Technology and Cyber Security, Logistics and Industrial Operations, Test and Evaluation, and Research and Engineering (see NAVAIR, "NAVAIR University," undated).

The example [name] gave is the best one I know of, of NAVAIR University, where they have additional training requirements above and beyond what DAU does, specific to their service. I think that they're really a complement to the DoD curriculum and I think that's okay for them to do. (DoD 4)

As an OJT example, the Air Force used Defense Acquisition Workforce Development Fund (DAWDF) funds to enable 15 contracting personnel to participate in the Air Staff OJT experiential program in FY 2017.[39] The Army also used DAWDF to fund internal rotational programs for career-broadening and interagency experiences, but more extensively, with over 800 of them—mainly in the contracting, program management, and engineering career fields.[40]

External T&D Options Currently in Use by the AWF

DoD and the individual services have also developed several options to provide acquisition T&D opportunities through nongovernmental organizations. Table 4.3 organizes them into the four categories of colleges and universities, industry (i.e., private-sector companies), professional associations, and commercial vendors.

Colleges and Universities

For a variety of reasons, including increasing flexibility and efficiency in providing acquisition training requirements, DAU has established a process that allows other organizations to offer courses, programs, or certifications that can be accepted as equivalent to one or more DAU courses.[41] Many colleges participate in this program; examples are the University of Virginia and the University of Maryland University College.

Table 4.3
Overview of Non-DoD Training Options

External Provider of T&D	Certification	Courses	Executive Education	Degree	OJT/ Rotations
Industry		X	X		X
Colleges and universities	X	X	X	X	
Associations	X	X			
Commercial vendors		X	X		

---

[39] HCI, 2018c, p. 18.

[40] HCI, *07-11-18 FY17 AWF Rotational Assignments DAWDF Year-in-Review Report*, spreadsheet received from HCI, 2018a.

[41] The process of approval for these courses is described in DAU Directive 708, *DAU Course Equivalency Program*, August 22, 2016: "An objective third party such as the American Council on Education (ACE) or other DAU-approved organization will serve as the reviewing body to make recommendations for approval of potential equivalent providers and their products."

All DAU 100- and 200-level courses (those generally required for Level I and Level II certification) are eligible for equivalency. More-advanced courses—for example, the "Understanding Industry" course (ACQ 315)—can be approved for equivalency on a case-by-case basis. CON 360, "Contracting for Decisionmakers," is one example and is a course we assessed in Chapter Two as contributing to the development of business acumen, knowledge of industry operations, or knowledge of industry motivation. About 10 colleges and universities provided 32 offerings of DAU-equivalent courses in FY 2017, with almost 900 students participating.[42]

Colleges and universities also confer degrees and provide executive education opportunities. In some cases, these college and university offerings are programs custom-designed for DoD acquisition personnel, but DoD personnel also have access to the broad set of offerings conferred by institutions across the United States. For example, one interviewee mentioned the Navy 810 program:

> [I]t's where an officer—and it's all for active-duty folks, it's not really open to civilians—but it's the active duty where the Navy will pay for that person to go and get their MBA degree, their master's. And Darden is one of the leading schools for that. Right now, I think we have 12 active-duty folks, male and female, that are in our MBA program. So their job for two years is to come to Darden and get their MBA and then once they graduate, they go back to the Navy to their command, they get an assignment. (Industry 13)

The Air Force's cooperation with the University of Tennessee in the Aerospace and Defense MBA program is an example of how working with educational institutions can assist with the areas of knowledge highlighted in Section 843. According to the Air Force's DACM website, this program is unique in that it is specifically designed to provide civilian aerospace industry professionals, including those in the Air Force AWF, with tools to understand the aerospace industry as a business.[43]

The Army also encourages advanced education at colleges and universities. The Army DACM operates the School of Choice program, which helps fund bachelor's and master's degrees in acquisition, business, or career-field disciplines. This program is funded using DAWDF money.[44]

Examples of executive education courses provided by colleges and universities are the two- to three-week "U.S. Navy Insights in Industry Management" course taught by the Darden School of Business at the University of Virginia (an Army version of the

---

[42] DAU, *Equivalency Provider Reporting FY 17* spreadsheet provided by DAU, July 2018a.

[43] Army, Navy, and Air Force personnel have participated in the ADMBA program in the past (see the 2015 graduating class at U.S. Army Acquisition Center, "The Aerospace and Defense MBA—Class of 2015," 2015a), but it was only mentioned by the Air Force DACM in our interviews. The Air Force participation number in Table 4.2 is based on the 2015 class of 35 students, of whom 18 were identified as Air Force.

[44] U.S. Army Acquisition Center, *Army Acquisition Education & Training (AET) Catalog 2015*, 2015b.

course is also taught) and a Navy "Understanding Defense Industry" course taught by the Kenan-Flagler School of Business at the University of North Carolina. The University of Tennessee also developed a four-day "Better Business Deals" course for the Air Force, which is oriented toward program managers and PEOs. For this course, faculty travel to different locations to help acquisition personnel learn more "about how industry thinks."[45]

Participation in the sample of available degree programs and executive education programs that were highlighted in our interviews and by DACMs is relatively limited, as shown in Table 4.4.

Table 4.4
Characteristics of, and Participation in, a Sample of University/College Degree and Executive Education Programs

Program	Provider	Service	Target Audience	Participation
Understanding Industry	Darden School of Business, University of Virginia	Army	Civilians, GS-12 to GS-15	23 to 24
School of Choice Program	Various	Army	Civilians, GS-11 and above	Not stated
Insights into Industry	Darden School of Business, University of Virginia	Navy	O-3/O-4; GS 9-11 (3-week course once a year) O-5/O-6, GS 14/15 (2-week course twice a year)	50 per course offering (approx.)
Understanding Defense Industry	Kenan-Flagler Business School, University of North Carolina, Chapel Hill	Navy	Program managers and deputy program managers	50 per course offering
Aerospace and Defense MBA Program	Haslam College of Business, University of Tennessee	Air Force	Civilians, GS-13 and above	18 (approx.)
Better Business Deals	University of Tennessee	Air Force	Program managers and PEOs	30 per offering

SOURCES: 2018 RAND Section 843 study interviews; U.S. Army Acquisition Center, 2015b; AF DACM website; December 6, 2018, e-mail from the Office of the Assistant Secretary of the Army for Acquisition, Logistics, and Technology.

---

[45] We learned from an interviewee that the contract for this course, which is being renegotiated, allows for up to eight presentations of the course. About 30 people attend each offering.

Industry (Defense and Non-Defense) Providers

Acquisition workforce personnel have opportunities to be exposed to industry operations and develop their business acumen outside the classroom in two important ways. The first is through rotations with industry—yearlong assignments with companies such as Amazon, Boeing, Lockheed Martin, and Microsoft, and others—to get hands-on experience with how industry works.[46]

The Army's "Training with Industry" program is exclusively for acquisition personnel, as is the Navy's "Training with Industry" program. The Air Force's "Education with Industry" program is advertised on the DACM website but allows participation by non-acquisition personnel,[47] and the Navy also has a "Tours with Industry" program that allows participation by officers and enlisted personnel who are not in an acquisition career field. The Secretary of Defense Executive Fellowship Program is a highly selective program for military and civilian personnel that provides an opportunity for senior personnel (not just from acquisition career fields) to work in industry for a year. Another DoD-sponsored program, but for civilians only, is the Information Technology Exchange Program (ITEP), which allows up to 50 personnel per year from any career field (but who must be working with information technology or cyber) to spend a year with industry.[48] Participating companies have included IBM, Microsoft, and Amazon Web Services.

Finally, the FY 2018 NDAA authorized DoD to introduce a public-private talent exchange that would allow the temporary assignment of civilian personnel to private-sector organizations and private-sector employees to DoD. Implementation guidance for this program was issued in July 2018, so there is not yet any information on companies that may participate or the number of DoD civilians who may apply. Participation in these rotational programs is competitive and limited, as shown in Table 4.5.

During interviews, we learned that companies such as Boeing, AT&T, Lockheed-Martin, and Deloitte offer their own industry "universities," and DoD personnel attend them on a limited basis. For example, one interviewee mentioned that DoD program managers in residence at his company for an industry rotational assignment will take courses at the in-house university during their time there. We also heard that DAU faculty members sit in on these courses as well:

---

[46] Industry participants in these programs were mentioned by DoD 2, DoD 3, and the program websites, such as Office of the Under Secretary for Personnel and Readiness, "SECDEF Executive Fellows," undated.

[47] A talking paper from Capt Hansen, who manages the training-with-industry program at AFIT, lists qualifications for participation; acquisition certification is not one of them.

[48] The DoD CIO website has a site for ITEP. "Cyber" was added to the law authorizing the program in 2016 (Pub. L. 114–328, Div. A, Title XI, §1123, Dec. 23, 2016, 130 Stat. 2455), so one interviewee referred to the program as CITEP. The amendment that added "cyber" also specified that no more than 50 people could participate per year.

Table 4.5
Target Audience and Participation Limits for Industry Rotation Opportunities

Program	Target Audience	Participation
Air Force Education with Industry	Captains and majors; civilian GS-11 through GS-15	37 per year for the past 20 years
Army Training with Industry	Acquisition personnel (Functional Area 51), grades 04 and 05	10–12
Navy Training with Industry	Supply corps officers only	3 in 2017
Navy Tours with Industry	Officers 03 and above; enlisted E-6 and above	30 in 2017
Cyber and Information Technology Exchange Program (CITEP)	Civilian GS-11 and above. Any career field, but must work with information technology	No more than 50 at one time
Secretary of Defense Executive Fellowship Program	Military O-5 and O-6; civilian GS-14 and GS-15	20 (sometimes more; four from each service)
Public-Private Talent Exchange	Civilian government personnel; not restricted to acquisition personnel	No limit stated in statutory guidance

SOURCES: Air Force: Air Force, Education with Industry Handbook, 2009, and Air Force, "Education with Industry Program Completes a Mid-Term Review," undated. Army: U.S. Army Acquisition Support Center (USAASC), email from official, July 17, 2018. Navy: Navy, "Navy Kicks Off New Tours with Industry Program," October 5, 2015, and NAVSUP Instruction 1520.7C, Training with Industry Program, March 9, 2017. CITEP: Chief Information Officer, U.S. Department of Defense, "DoD CITEP Frequently Asked Questions," undated. Secretary of Defense Executive Fellowship Program: Office of the Under Secretary for Personnel and Readiness, undated, and Public-Private Talent Exchange, July 19, 2018, memorandum from Deputy Secretary of Defense.

NOTE: For the Public-Private Talent Exchange, in an October 15, 2018, telephone discussion with functional representatives, we were told that HCI has nominated five people for participation in the Public-Private Talent Exchange program and that nine companies have expressed interest in participating.

We had an opportunity to attend the Boeing Leadership School, their leadership academy near St. Louis, and that was fantastic. We here at DAU also have opportunities to attend those schools and others, and we try to take advantage of that whenever we can. The University of North Carolina or Boeing Training Academy or Lockheed Martin Leadership School or Raytheon, we go to all those. So we want to know what defense industry is using. (DoD)[49]

DAU sends probably four [DAU] faculty a year to go and find out what they're teaching and they're pretty open. As a matter of fact, in one instance, we got a brief on their [named working group], which is their independent lobby organizations, and they gave us a very insider view. (DoD 6)

---

[49] Numeric identifier omitted to protect interviewee confidentiality.

Professional Associations

A number of professional societies, such as the NCMA, the PMI, and several others that provide a variety of professional certifications, were mentioned by interviewees as potential sources of knowledge about business acumen, industry operations, and industry motivation—though we will see below that opinions about the value of some certifications were mixed. Professional societies have also participated in the DAU course-equivalency program.

Commercial Vendors

Commercial vendors (Booz-Allen is one example) participate in the DAU course-equivalency program, so they are another source of the type of knowledge mentioned in Section 843. Participation in this activity is widespread; some 25 commercial vendors offered 676 DAU-equivalent courses in FY 2017, with over 13,000 students (1,240 of them from DoD) participating.[50]

Resources Supporting the Use of External T&D
Advertising

The availability of many external resources for acquisition workforce T&D is advertised in several places. The DAU website includes information on all commercial, industry, and college providers of DAU-equivalent courses; the Army, Air Force, and Navy DACM websites provide information on their industry rotation programs; and each DACM advertises degree-granting programs available for AWF personnel.

The Navy DACM has also made an effort to increase awareness of the Secretary of Defense Executive Fellowship program and the CITEP program—descriptions of the programs and information about how to apply were discussed in the January–March 2018 quarterly "DACM Corner" publication.

Funding

DAWDF has facilitated a wide array of T&D activities. The Army, Navy, and Air Force have all created civilian opportunities for AWF external education and training at approved universities and colleges by using DAWDF for tuition assistance. All three services have also tapped DAWDF for student loan repayment programs—though the repayment programs do not necessarily apply to acquisition-related education.[51] Military and veteran AWF personnel also have access to funding for education through the Tuition Assistance and Post-9/11 GI Bill programs. In addition, as noted earlier, the Air Force used DAWDF to cover the costs for 15 contracting professionals partici-

---

[50] DAU, 2018.

[51] The use of DAWDF for loan repayments for all three services is mentioned in the *Department of Defense Acquisition Workforce Development Fund: 2017 Year-in-Review Report.* This report also mentions the use of DAWDF for tuition assistance for civilians in the Navy and Air Force. Army use of DAWDF for tuition assistance is mentioned on the Army DACM website (U.S. Army, "DAWDF Program," 2018).

pating in the Air Staff OJT experiential program in FY 2017, and the Army also used DAWDF to fund internal rotational programs for career-broadening and interagency experiences on a more extensive basis. Finally, DAWDF has supported broader investments, such as the AWQI e-workbooks we analyzed earlier in this report and investment in a centralized database:

> We also use the DAWDF for our system, for our talent management system which we mainly use for compliance for DAWIA, for tracking of the certification requirements, continuous learning requirements and Acquisition Corps. (DoD 1)

While DAWDF has been used extensively to foster AWF T&D efforts, it is important to note that DoD has typically not fully executed DAWDF funds. This appears to be improving: In early 2018, HCI reported that the FY 2016 DAWDF program was executed at 92 percent, which was the "best ever" in DAWDF history,[52] and in a September 2018 Workforce Management Group (WMG) briefing, HCI shared that all DAWDF users expected to execute 100 percent of funding.[53]

Policies

In a March 2018 memo entitled "Engaging with Industry," the Deputy Secretary of Defense wrote:

> Our National Defense Strategy (NDS) directs our intentional engagement with industry to harness and protect the National Security Innovation Base as well as modernize key capabilities. Cultivating a competitive mindset requires that we optimize our relationships with industry to drive higher performance while always remaining within the letter and spirit of ethics and procurement regulations. . . . Proactive engagement will maximize support to the Warfighter; set realistic expectations and technologically achievable requirements; enhance the ability of organizations to meet cost, schedule, and performance objectives; and establish policies and business practices that promote the long-term viability and competitiveness of the industrial base supporting defense.[54]

The memo encourages the types of engagement that could lead to DoD-industry rotations or other training opportunities for AWF personnel to learn more about business acumen, industry operations, and industry motivation, as well as opportunities for industry personnel to participate in DoD AWF-focused training. The memo also seems to have been a catalyst for "reverse industry days," relatively short events in which industry professionals share their perspectives with defense acquisition per-

---

[52] HCI, 2018c.

[53] HCI, "A&S Human Capital Initiatives," updates presented to Workforce Management Group (WMG), September 5, 2018d, briefing provided to RAND study team by WMG group member on October 15, 2018.

[54] P. M. Shanahan, Deputy Secretary of Defense, "Engaging with Industry," memorandum, March 2, 2018a.

sonnel. While reverse industry days are more of an opportunity to share experiences than to participate in training, they are another proactive form of engagement with industry—one that may also informally help to confer Section 843–related knowledge. As one interviewee put it:

> We have industry panels, we have government people attend. So you have industry personnel sitting up there talking about some of the challenges that they have with government. I don't want to say it opens our eyes, but I think it gains a little more clarity in the workforce; members can gain that insight into some of the decisions they make, the negative impacts that it has on industry. I think you have all of government looking at ways to better understand industry. (Industry 1)

We learned through our interviews and subsequent correspondence with acquisition workforce career management officials that defense acquisition workforce personnel have participated in this form of public-private exchange as well. For example, in November 2018, the National Geospatial-Intelligence Agency (NGA) hosted a half-day reverse industry session, with agenda items such as how industry makes decisions on whether to bid for NGA work, how industry develops its strategies, and insights from former federal employees currently working in industry.

Determining Whether Greater Use of External T&D Is Needed to Fill Gaps

As described in previous section, the evidence suggests that DoD has been using a range of external T&D resources to build business acumen, knowledge of industry operations, and knowledge of industry motivation. The current capacity of these externally provided programs is quite limited, yet DoD also incorporated external T&D resources into DAU training to reach a broader set of personnel. This section addresses the question of whether DoD should be using more external T&D resources to address the gaps identified in Chapter Three. We start by summarizing findings from interviews about the perspectives of DoD and industry leaders regarding greater use of external T&D. We then describe some additional efforts DoD might engage in to conduct a more rigorous needs analysis and assess internal and external T&D options against a common set of criteria. We conclude with some additional evidence from interviewees regarding external T&D and what it contributes to DoD's portfolio of T&D offerings.

Interviewee Perspectives on Whether More External T&D Is Needed

We asked DoD leadership and industry interviewees whether they believed that DoD should be using more external T&D resources, and the majority of respondents suggested that more external T&D would be a good thing. Table 4.6 provides some

Table 4.6
Perspectives on Whether to Increase Use of External T&D

Type of External T&D	Interviewee Quotes
Industry rotations and exchange programs	"Again, the one area that I'd like to participate in a lot more or participate in is the Training with Industry. Because I think there are—you know, we don't have that balance that in some cases we could get from training with industry." (DoD 14)
	"So, I don't think we necessarily send enough of our guys out to software companies. . . . I'm fairly confident we could be sending more. . . . We need to spread out because there's a lot of other areas, everything from IT to the Logistics aspects that we need to take into account." (DoD 11)
Executive education and degree programs	"Provide funding and incentives for DoD Acquisition personnel to complete business school courses. It's kind of like an education with industry but it's actually sending people away to get an advanced degree. . . . You should do something similar for people with advanced business degrees or maybe something less than an Executive MBA but yeah, give them a better understanding, knowledge of how industry works and how you can use that to the advantage of government." (DoD 21)
	"I know many of the companies offer internal training . . . the Lockheeds, the Boeings, the Raytheons, Northrup Grummans—they all have their own internal corporate universities which they potentially could or do offer the opportunity for government personnel to come take some of that training and get some exposure. That would be really good." (DoD 8)
Professional certifications	"I believe leveraging industry and/or at least working closely with industry training providers, et cetera, to deal with, agree on, sort of like I said, this consistent industry-wide standard and set of training standards would allow the DoD to utilize its resources a little bit differently and focus on the unique stuff while utilizing sort of this common base of industry stuff to do sort of what's really foundational and core learning." (Industry 7)
	"They go, 'Hey, we offer some training like DAU and people value our certificate.' They make money doing that and our answer is, 'Well, we have our own certification. So ours is good enough.'" (DoD 7)
Commercial training courses	"We're also looking at . . . In Learning. Linked In Learning has 6,000 courses online available 24/7 for 60,000 people in the Acquisition Workforce. . . . It's $7 a head. Seven dollars per person for one year for 24/7 access to 6,000 courses that range from agile to cloud to PMP to programming an iPhone to Logistics, Supply Chain Management. They've actually mapped all those courses to the SES ECQs. So I was sold on it." (DoD 2)
Incorporation of external resources into DAU T&D	"Can more be done? Absolutely. And that's one of the things that these last three days with those four DAU representatives in that we very specifically talked about how we can improve our—we have an actual memorandum of understanding of how we are going to talk to one another and how often and what sorts of resources are going to be applied." (Industry 9)
	"DAU used to have a lot of industry participation in its courses. I mean, that was a steady rising line up to 2010 and since 2010, it's dropped like a rock for a couple of reasons. But I think the major reason has been that in the era of sequestration and the industry trying to protect its earnings per share . . . . So people in the industry just don't have the time or the luxury to go to like a 12-week Program Management course." (DoD 6)

SOURCE: 2018 RAND Section 843 study interviews.

examples of how interviewees described the need for more external T&D resources to address gaps in Section 843–related knowledge.

Perspectives of interviewees on whether external T&D should be used more varied according to type of training. Key takeaways by type of training include the following:

- **Industry rotations:** Interviewees who discussed this form of development tended to suggest that opportunities for industry rotations should be increased and placed a particular emphasis on the need to increase opportunities for civilians—though many interviewees also noted substantial barriers to scaling these rotation programs. Efforts to expand civilian industry rotations through the Public-Private Talent Exchange program suggest that DoD is already moving toward expansion of these opportunities.

- **Executive education and degree programs:** Perspectives on executive education and degree programs from universities and colleges were largely in favor of increased use. Some interviewees advocated for the greater use of these opportunities, including both general degree and executive education programs and tailored programs that are designed in collaboration with DoD. Several providers of these programs that are currently in use by DoD suggested that they could increase capacity to accommodate more DoD personnel. Some interviewees also suggested that it could be useful to increase participation in executive education programs offered by corporate universities like those offered by Boeing and Lockheed Martin. However, some interviewees reported that the current use of executive education and degree programs was sufficient.

- **Professional certifications:** Perspectives on professional certifications and other interactions with associations were mixed, with the DoD interviewees who discussed them mostly reporting use to be sufficient, while some industry stakeholders suggested that much more should be done to collaborate with professional associations. Interviewees suggested that a closer look at how professional and DAWIA certifications could be used in complementary ways may be warranted.

- **Courses from vendors:** While one interviewee suggested that DoD could benefit from looking into Coursera and understanding whether this type of platform might offer value to DoD, there was otherwise little discussion among interviewees about the need to increase use of external courses from commercial vendors. The reasons for this may have been twofold, including greater overlap with what is already provided internally by DAU and the limited incorporation of experiential and applied learning in these types of courses (something that interviewees argued is an advantage of many external T&D options currently in use to confer business acumen and knowledge of industry).

- **Use of external resources for DAU courses:** Perspectives on the incorporation of external resources into DAU training were also mixed. Some interviewees felt that DAU was doing a lot of this already and that these efforts were sufficient, including the use of guest speakers, the invitation of industry participants to DAU courses, and regular efforts of faculty to interface with professional associations and private-sector companies. However, others argued that there was always more that could be done to enhance DAU's use of external resources. For example, one interviewee noted that participation of industry personnel in DAU courses had declined in recent years for various reasons, such as sequestration and lesser interest among private-sector organizations, and that it would ideal to increase the level of participation. Actual industry course participation data provided by DAU suggest that participant rates have fluctuated rather than steadily declined: The number of industry graduates of DAU courses decreased from 306 in FY 2016 to 181 in FY 2017, but it then increased to 238 in FY 2018.[55]

A Rigorous Approach to Determining Whether More External T&D Is Needed

While interviewee perspectives were informative to determining whether more external T&D was needed to fill gaps, they were not sufficient. In this section, we discuss two areas in which additional analysis could be conducted: an enhanced training needs analysis and a rigorous assessment of DoD's full portfolio of T&D offerings.

Conducting a More Robust and Targeted Needs Assessment

The literature on developing effective training suggests that a needs analysis is an important best practice. The strategies for conducting needs analysis vary in the literature, but there is agreement that the goals of needs analysis are to determine "what needs to be trained, for whom, and within what type of organizational system."[56] DoD efforts around competency models and proficiency assessments are valuable endeavors in identifying what needs to be trained and for whom. However, there may need to be a more targeted focus on the areas identified in Section 843 and an explicit effort to build in competencies and assess gaps where relevant. As noted earlier in the report, our efforts to identify Section 843–related knowledge gaps relied on competency models and proficiency assessments that may not have sufficiently examined needs for knowledge in the areas of business acumen, industry operations, and industry motivation. Systematic efforts to build in related competencies, necessary levels of proficiency in each, and more robust and targeted assessments of gaps may be in order.

Such endeavors would also take into consideration the current state of T&D completed by the AWF, to include work and education experiences obtained prior to joining the AWF. However, we learned that while total participation numbers in different T&D activities (e.g., number of students in a particular course) are tracked, individual personnel records do not routinely capture all T&D offerings that a specific acquisition

---

[55]   DAU data provided on January 3, 2019.

[56]   E. Salas, S. I. Tannenbaum, K. Kraiger, and K. A. Smith-Jentsch, "The Science of Training and Development in Organizations: What Matters in Practice," *Psychological Science in the Public Interest*, Vol. 13, No. 2, 2012, pp. 74–101.

professional has taken. In a related vein, DoD's primary database for civilian personnel management, the Defense Civilian Personnel Data System (DCPDS), is missing potentially useful data for a significant portion of the AWF, such as the types of graduate degrees that AWF members hold.

Systematically Assessing the Portfolio of T&D Options

Many interviewees described internal and external T&D as both playing important and complementary roles in addressing the needs for Section 843–related knowledge. As one interviewee described:

> I think they're very complementary. . . . They're not competing with us. They're complementing us in terms of capabilities or expertise. (DoD 8)

This suggests that an optimal portfolio of T&D may include both internal and external options and that the task for DoD is to determine the appropriate mix of offerings. After identifying the specific knowledge requirements and understanding which members of the AWF face gaps in Section 843–related knowledge, the next step is to identify the type and provider of T&D that will be most appropriate for addressing these knowledge requirements. Determining the appropriate type and provider of training requires systematic comparison of options across DoD's portfolio of internal and external T&D offerings according to a common set of criteria.

We identified four categories of criteria that DoD might want to consider in determining the appropriate way to fill gaps through its assessment of external and internal T&D offerings (Table 4.7). Because external and internal T&D are often funded and managed in different ways and may be subject to different types of legal and regulatory constraints, it is important to consider how various T&D options stack up against criteria in these areas. For example, DAWIA requirements may affect the need for standardization and the flexibility in providing training. A second set of criteria relate to the design and delivery of T&D. DoD might want to examine a wide range of aspects of design and delivery, from content to structure to participants. All T&D requires resources, and these resources are often constrained, so resources must be considered in evaluating various T&D offerings. Finally, the outcomes related to participation in T&D and evidence of effectiveness are critical to consider in assessing the need for more external T&D.

While the examples of possible criteria are laid out in a general way in Table 4.7, these may need to be converted into more-specific expectations about what is needed to address a particular knowledge requirement. Many resources can help to inform the criteria used to assess T&D offerings, including the literature on best practices, perspectives of key stakeholders within DoD, and reflection or formal evaluation to determine what has been most effective for similar types of T&D. One area where best practices from the literature might be particularly useful is with regard to best practices for the design and delivery of effective training. We conducted a targeted review of the

Table 4.7
Examples of Criteria That Might Be Used to Assess T&D Options

Category of Criteria	Criteria
Governance and regulation	• Management and oversight • Legal and regulatory constraints • Funding source
Design and delivery	• Learning objectives • Structure • Content • Pedagogy • Participants
Resources required	• Capacity • Cost • Time away from work
Outcomes and effectiveness	• Participant reactions • Knowledge learned • Changes in behavior • Impacts on the organization

SOURCES: Examples of factors that might be important to organizations in deciding between different T&D options were drawn from the literature and interviews with DoD and industry experts. Outcome and effectiveness criteria are drawn from the Kirkpatrick Model, described in greater detail in Chapter Five.

literature on best practices for T&D and identified eight characteristics of high-quality training (Table 4.8). These best practices could be modified into criteria for design and delivery. In other words, rather than the more general "delivery," a criterion could be "degree to which applied and experiential learning are used."

Additional Evidence to Inform External T&D Decisions

While it was outside the scope of the study to conduct a full assessment comparing internal and external T&D along the criteria laid out above, evidence from our interviews provided some insights into how external and internal T&D offerings might compare in a few areas. These additional findings are summarized in Table 4.9, and we provide a description of the interview evidence below.

Exposure to Experts in the Field

According to Harvard Business Publishing (2016), "learners construct knowledge through interactions with those who know more than they do or those who have different experience and perspectives—whether a peer, coach, boss, expert, or facilitator." This underlines the importance of interacting with those who have expertise in Section 843–related knowledge to ensure high-quality T&D. Some interviewees raised concerns that internal T&D might not be as appropriate for T&D that intends to offer the industry perspective. According to one interviewee:

Table 4.8
Best Practices and Possible Criteria for Assessing T&D

Best Practices (Possible Criteria)	Summary of the Literature
Aligned with the organizational mission and goals	Studies suggest that high-quality T&D is designed to be closely aligned with the mission and goals of an organization. This ensures that it feels meaningful to employees, has buy-in from leadership as a priority, and is more likely to translate directly into organizational effectiveness.[a]
Clearly outlined learning objectives	Studies suggest that it is important to ensure a clear understanding of the objectives of T&D and what the specific aim of the training is in terms of behavioral change.[b]
Tailored to the specific work context and prior experiences	Several studies found that a close relationship between training content and work tasks increases the likelihood that training leads to changes in behavior. The ability for participants to connect training to prior knowledge is also important and, according to the literature, improves outcomes.[c]
Opportunities for applied, action-oriented, and experiential learning	Studies frequently emphasized the opportunity to apply what is learned as important and described "experiential learning" as being an important T&D approach to ensure application of learning. The literature described that this learning can be achieved through on-the-job practice or other types of training content that apply learning, such as case studies and projects. Ensuring that training is delivered close to the time it needs to be used on the job is another way for ensuring application of learning.[d]
Offered as a process with opportunities for reinforcement over time	Studies suggest that training is most effectively transferred to practice when it is developed as a process rather than a one-time thing. In this process, opportunities for application and feedback follow the initial training activity. Reinforcement by supervisors was described as also being important, and linking of T&D participation to performance reviews was described as one way of doing this.[e]
Involves active and collaborative learning	Studies described the need for participants to play an active role in the experience and engage in collaborative experience with other trainees or mentors. According to the literature, interactions with those who have different experience and perspectives can be particularly valuable.[f]
Is accessible to learners and organizations	Several resources mentioned that training is made accessible to learners in various ways. For example, some argue that training should be short to avoid pulling individuals away from jobs and that employers should set aside time and money to explicitly support T&D. Studies also emphasized the value of mobile learning that can be accessed in various locations and keeping training accessible in various ways.[g]
Incorporates technology to support application and learning as process	While the evidence on face-to-face instruction versus online learning is mixed, it suggests that online and blended learning can be effective if used in the right ways—i.e., to support some of the other effective practices highlighted in this table. For example, technology can facilitate more applied learning opportunities, can help to deliver training as a process over time, and can support the accessibility of training.[h]

Table 4.8—continued

Best Practices (Possible Criteria)	Summary of the Literature
Considers learner mindsets and emotions	Studies indicate that the mindsets of learners are important and that ensuring self-efficacy and motivation is important to the success of training. T&D providers are increasingly focusing on content that not only addresses specific training content but also builds confidence and motivation among participants that what is learned can be used to improve work practices and business outcomes.[i]

[a] Salas et al., 2012; H. B. Bernhard and C. A. Ingols, "Six Lessons for the Corporate Classroom," *Harvard Business Review*, 1988; OPM, 2012; Harvard Business Publishing, *Accelerate Leadership Development with Optimal Design: Six Key Principles*, 2016; M. Weinstein, "The Bottom Line on Leadership," *Training*, 2012, pp. 49–52; M. Beer, M. Finnstrom, and D. Schrader, *The Great Training Robbery*, Cambridge, Mass.: Harvard Business School, Working Paper 16-121, 2016.

[b] Salas et al., 2012; L. A. Burke and H. M. Hutchins, "Training Transfer: An Integrative Literature Review," *Human Resource Development Review*, Vol. 6, No. 3, 2007, pp. 263–296.

[c] Burke and Hutchins, 2007; Association for Talent Development, *The Science of Learning: Key Strategies for Designing and Delivering Training*, Alexandria, Va., 2017.

[d] Filipkowski, 2014; OPM, 2012; Harvard Business Publishing, 2016; Stanton and Stanton, 2017; Weinstein, 2012; Salas et al., 2012; Association for Talent Development, 2016a; P. Gurdjian, T. Halbeisen, and K. Lane, "Why Leadership-Development Programs Fail," *McKinsey Quarterly*, January 2014; A. Hughes, "How Elearning Benefits Corporate Leadership Training: Bridging the Gap," *Leadership Excellence Essentials*, Vol. 35, No. 3, 2018, pp. 16–17; R. Grossman and E. Salas, "The Transfer of Training: What Really Matters," *International Journal of Training and Development*, Vol. 15, No. 2, 2011, pp. 103–120; L. A. Burke and H. M. Hutchins, "A Study of Best Practices in Training Transfer and Proposed Model of Transfer," *Human Resource Development Quarterly*, Vol. 19, No. 2, 2008, pp. 107–128.

[e] Weinstein, 2012; Stanton and Stanton, 2017; OPM, 2012; Harvard Business Publishing, 2016; Burke and Hutchins, 2007; Salas et al., 2012; Bernhard and Ingols, 1988.

[f] Burke and Hutchins, 2007; Harvard Business Publishing, 2016; Salas et al., 2012; Stanton and Stanton, 2017; P. Donnithorne-Nicholls, "How to Engage Modern Learners," *Human Resources Magazine*, Vol. 22, No. 1, 2017, pp. 8–9.

[g] Beer, Finnstrom, and Schrader, 2016; Donnithorne-Nicholls, 2017.

[h] OPM, 2012; Hughes, 2018, Donnithorne-Nicholls, 2017; M. T. Jacot, J. Noren, and Z. L. Berge, "The Flipped Classroom in Training and Development: Fad or the Future?" *Performance Improvement*, Vol. 53, No. 9, 2014, pp. 23–28; Salas et al., 2012; Burke and Hutchins, 2007; A. Rio, "The Future of the Corporate University," *Chief Learning Officer*, Vol. 17, No. 4, 2018, pp. 36–56.

[i] Association for Talent Development, 2017; Beer, Finnstrom, and Schrader, 2016; Gurdjian, Halbeisen, and Lane, 2014; Harvard Business Publishing, 2016; Salas et al., 2012; Burke and Hutchins, 2007; Weinstein, 2012.

The perspective they're very often going to receive within DoD is the very same thing that they already have. That outside perspective is crucial to their development. (Industry 15)

According to interviewees, internal T&D resources can ensure industry expertise both by drawing on instructors with deep knowledge of industry and by including industry experts as participants in T&D alongside DoD personnel. One interviewee described the benefits of having industry participants in DAU courses:

Table 4.9
Perspectives on External T&D Related to Possible Criteria

Criterion	How External T&D Compares with Internal T&D
Exposure to experts in the field	Some interviewees perceived that external T&D provided greater exposure to trainers and participants with expertise on industry.
Frequent opportunities to apply learning	Some interviewees perceived that external T&D offerings provided many opportunities for experiential learning, while internal T&D was reported to be largely course based (though it did aim to incorporate experiential components).
Tailored to specific work context	Some interviewees reported that a strong benefit of internal T&D is its strong connection with the work context within DoD and prior knowledge of DoD personnel, whereas external T&D varied in the degree to which it emphasized connections to the DoD context.
Immersive environment	Some interviewees perceived that the immersive environment in some external T&D offerings offers an opportunity for more learning relative to T&D taken while on the job.
Time away from work	Some interviewees reported that the longer length of some external T&D, like rotations and degree programs, led to greater workforce reductions relative to shorter internal and external T&D opportunities.
Capacity	Some interviewees highlighted the small capacity of tailored forms of external T&D, while capacity for participation in general courses and degree programs was much greater. DAU also faced some capacity constraints, but not as great as those for rotations and tailored educational programs.
Cost	Some interviewees reported that the costs of external T&D were higher, especially for tailored forms of T&D, such as rotations and custom executive education programs.
Funding stability	Some interviewees highlighted uncertainty regarding funding for external T&D options.
Legal and regulatory restrictions	Some interviewees highlighted concerns about legal or regulatory requirements, including conflict of interest requirements and the ability to backfill positions.

We very much appreciate having the industry partners in the class with us. Because then the conversations are different and you get different viewpoints and background information when you're doing the instruction or the intact team training, if you will. (DoD 7)

Frequent Opportunities to Apply Learning

The literature on best practices for T&D emphasizes the importance of applied learning. According to OPM's 2012 *Executive Development Best Practices Guide*:

Traditional classroom settings have given way to a much more hands-on, action-oriented approach to development. Executives derive more value from this type of learning by directly applying knowledge to real problems and situations.[57]

---

[57] OPM, 2012.

One study found that high-performance organizations are nearly three times more likely than lower-performing firms to use experiential learning for development of both senior-level and frontline leaders.[58]

When describing the benefits of external T&D opportunities, interviewees emphasized the strong focus on experiential learning and the many opportunities to apply what is being learned about interaction with industry and demonstration of business acumen. According to one interviewee:

> I think [there is] no experience [that] beats hands-on experience, you know. Nothing beats—even training, it's better than training. So even when you say going to formal education, nothing beats working with industry directly or working from an industry perspective. So having industry experience would probably be the best way to have that first-hand experience. (DoD 19)

Because many of the internal T&D offerings that addressed business acumen. knowledge of industry operations, and knowledge of industry motivation were course based, participation in rotations and some executive education offered T&D opportunities that were perceived to offer greater exposure to applied learning. One interviewee discussed this comparison:

> Well, it's a big difference between book knowledge and actually seeing it in action, right? We can take earned value and we can understand actual cost for the work performed and the budget cost for the work performed. But it's a whole different thing—and that's the kind of thing DAU will teach you, they'll teach you how to—you get the gold card and you learn how to crunch the numbers. But when you do it in industry, you're really making judgements, you're really holding people accountable. (DoD 18)

Tailored to Specific Work Context

The literature suggests that learners must be able to connect what is learned back to the environment in which it needs to be applied. Burke and Hutchins emphasized the importance of content relevance, or a "close relationship between training content and work tasks."[59] ATD describes the importance of connecting learning to prior knowledge when designing or delivering training.[60] This suggests that T&D offered to DoD personnel needs to make clear connections back to the DoD context and the work tasks on which personnel will apply knowledge. The explicit focus of DAU and other internal T&D providers is to tailor T&D for DoD personnel. External T&D, on the other hand, may vary in the degree to which it makes connections to the DoD context

---

[58]   Association for Talent Development, 2016a.

[59]   Burke and Hutchins, 2007.

[60]   Association for Talent Development, 2017.

and the work activities of DoD personnel. One interviewee described these limitations of some external T&D offerings:

> I know coming up in my career for financial management, I took classes at the graduate school which taught financial management. It taught federal budgeting, it taught accounting and those sort of things. But when you get into DoD, it did not teach the uniqueness of how do we . . . develop budget justification material within DoD. (DoD 19)

Interviewees often placed an emphasis on custom-designed T&D when discussing valuable sources of knowledge for Section 843–related knowledge and acknowledged the important role that DoD can play in helping to inform the design and reinforcement of external T&D. According to one interviewee:

> Just recently, we had external non-DoD training solutions . . . being considered to ensure the best training solutions are offered to the students. External forces often need to be enhanced with DAU instruction to ensure applicability to the workforce members. (DoD 7)

Yet, despite acknowledgement of the important role that DoD can play in helping to ensure that external T&D is tailored to the DoD context and that learning is reinforced in the DoD context, the efforts to make these linkages appeared to be limited for many types of external T&D.

Immersive Environment

Some interviewees described the value of immersing DoD personnel in an industry-focused environment and of separating personnel from their day-to-day work in developing business acumen and knowledge of industry. According to one interviewee:

> It's most effective when you can get somebody out of that environment and so, they're not trying to take an online course while they're in the whirlwind of daily operations and trying to split their attention, you know? So if you can get them out of that environment and get them into more of a training environment or immersion with a company, that's so much better. (DoD 21)

Interviewees suggested that external T&D opportunities might offer greater opportunities for this immersive experience because they were provided offsite and required dedicated attention to the T&D. Some also observed that immersion in private-sector culture through rotations can expose individuals to the faster pace of industry that might help to open up the eyes of DoD personnel to other ways of doing business. As one interviewee described it:

It's just, you know, a different environment, a different culture, the pace of maybe how things are done. . . . I think it brings them a learning that they wouldn't have exposure to while on active duty or at a service school or whatever. (Industry 12)

However, it is important to note that there is tension between immersion in an industry environment and the ability of individuals to make connections to prior knowledge and the context in which knowledge will be applied, which was described previously as being important. This suggests that while immersion in an industry environment might help to connect the knowledge of industry to job tasks conducted in an industry environment, there might be concerns about the ability to transfer this knowledge back to the DoD environment if sufficient connections are not made as a part of the external T&D.

Time Away from Work
One of the most commonly mentioned barriers to use of external T&D was a lack of time. According to one interviewee:

I think the biggest thing is that we're all wicked busy. . . . Not a whole lot of people can afford to take two or three weeks off from their primary job to go off to take training. (DoD 10)

Participation in any type of T&D, internal or external, creates disruption to workflow as personnel step away from work duties. Yet, because some forms of external forms of T&D, such as industry rotations and degree programs, required weeks or months of immersive participation and were potentially less flexible in terms of when participation was possible, workforce disruptions were a particular concern. One interviewee described this as being a barrier to ensuring that the most productive individuals were provided with external T&D opportunities:

You don't get your best—because you're going to lose that individual from an assignment for a period of time—a month, three months, a year—and so you're not going to get your best performers off the street for a year while they're out in industry. (Industry 2)

DoD has explored some approaches to addressing workflow disruptions for industry rotations, including efforts to facilitate backfilling of positions and efforts to create exchange programs that fill gaps with industry personnel.

Capacity
There was variation in the capacity of external T&D opportunities. Opportunities to take courses and enroll in occupational certification and degree programs at colleges and universities were open to a broad set of institutions across the United States, and there is open access and unlimited capacity for professional certifications and many

vendor-offered courses. However, the capacity in the tailored DoD-specific T&D opportunities that interviewees tended to emphasize as being important for developing Section 843–related knowledge, such as industry rotations and tailored executive education programs, was limited. One interviewee described this limited capacity:

> The one thing we wondered was even if DoD can do more in industry, there's a capacity issue of yeah, how many—because the industry training programs that we are aware of, they're pretty small numbers of people that go through them. (DoD 23)

Further, there was limited opportunity to expand capacity substantially in these programs, according to some interviewees. Although private-sector companies may have the capacity to host more DoD fellows than they do now, the overall number of DoD personnel at a company at any given time is likely to be small.

> I mean, as long as we have opportunity within the company to provide, you know, a productive term while they're here and they're receiving what they want to receive from the fellowship then, you know, I think we could host, you know, quite a few more folks than we're hosting today. But, I mean, I'm sure at some point there would be some sort of saturation, but we could definitely grow the program. (Industry 12)

Moreover, there is a limited number of companies that DoD may want to or be able to tap for these opportunities, so the overall number of rotation opportunities is always likely to remain small.

On the other hand, internal overall DAU capacity is high—counting both classroom and online offerings, there were more than 900,000 completions of individual DAU courses in FY 2017.[61] However, capacity for some individual courses related to business acumen, knowledge of industry operations, and knowledge of industry motivation is limited: As noted in Table 4.1, there were close to 1,300 graduates of ACQ 315 in FY 2018, and, according to one of our interviewees, this is close to the limit:

> And 315 is a popular class at DAU. It's a requirement for Program Management. It's a pick list choice for Contracting. I just pulled up our data and **we are at capacity in FY18 for ACQ 315**. (DoD 1—emphasis added)

---

[61] *RAND Industry Grad Request FY17* spreadsheet provided by DAU, August 2018. This number means, for example, that if 100 people completed course A and the same 100 people completed course B, there would be 200 course completions.

Cost

The cost of T&D is always an important consideration when determining what DoD personnel are offered. One interviewee described these cost-related concerns, noting the important role of direct costs and indirect costs related to covering work duties:

> It would be great if you could have half your workforce in some kind of training or education at any one time. But there's a lot of cost involved with that. . . . It's not just paying for the education but, you know, now you have to basically have two positions because somebody has got to be in the desk doing the work. (DoD 7)

While it was outside of the scope of the study to analyze the costs of various types of T&D, some interviewees noted the high cost of external executive education and degree programs offered by colleges and universities as a barrier to their use. For example. one interviewee discussed issues related to covering work duties:

> Unless you are kind of hitching your wagon to something that exists at a university that's a pay-as-you-go, they can be very expensive to put in place, tens of millions of dollars a year, depending on how much you want a custom program. (DoD 18)

In addition, while industry rotations and participation in corporate university education programs were often provided without cost to DoD personnel, there were other costs associated with industry rotations. For example, these programs required personnel to travel and potentially even relocate temporarily, and the length of rotations sometimes required positions to be backfilled while individuals were participating. For these higher-cost external T&D options, the value and effectiveness must be sufficiently high to ensure positive return on investment (ROI) and justify the use of these resources over internal T&D.

Funding Stability

The stability of funding is important for maintaining T&D programs, and because internal and external T&D was often managed and funded separately, the stability of funding may also differ. While a number of interviewees were grateful for the funding provided through DAWDF, Tuition Assistance, and other pockets of funding dedicated to external T&D, some interviewees suggested that the funds used for T&D were at great risk of being cut at any time, leading to a hesitancy to invest resources in building out these resources. As described by two interviewees:

> [W]hen funding cuts come, most of the time training and travel are the first two things that get cut. (DoD 19)

> Training is the first area that's cut [in DoD]. If there are any discretionary funds available, it's the training stuff that they serve up before they have to cut people. (Industry 2)

Legal and Regulatory Restrictions

Both internal and external T&D within DoD may be subject to various legal and regulatory requirements, though these differ for different types of T&D and different providers. According to some interviewees, these restrictions are greater for external T&D. Some interviewees expressed concerns about legal or proprietary issues that might arise whenever DoD is interacting with industry outside of regular acquisitions processes. For example, according to one interviewee:

> I think it's a great opportunity but when it's civilians, at least when we tried to do it, we ran into too many legal restrictions and legal problems with going to work in industry. (DoD 14)

With regard to regulations, one statutory restriction for the Public-Private Talent Exchange program affirmed by the implementation guidance[62] is that it does not allow DoD positions vacated by participants in such an exchange to be filled by someone else:

> Some of the challenges with that authority is that one, you can't backfill the position, you can't outsource the work. It's supposed to be transparent from a cost perspective, which is kind of unrealistic and difficult, because work still needs to be done. (DoD 26)

The leadership of HCI recognizes that, while the authorization for the exchange is an excellent opportunity for civilian AWF personnel, the restriction may make organizations reluctant to participate, so it has recommended that the FY 2020 NDAA include an amendment to the Public-Private Talent Exchange authority established in FY 2017 NDAA Section 1104 that removes the backfill prohibition and further incentivizes industry participation by allowing it to bill indirect costs of the exchanges.[63]

Summary

DoD has many different options for meeting knowledge requirements related to business acumen, industry motivation, and industry operations, including courses, executive education programs, degrees, certifications, rotations, and OJT. DoD Currently offers a range of T&D options from both internal and external sources that are intended to address these knowledge requirements. Internally, DAU courses and on-the-job experience were the most commonly mentioned sources of T&D addressing these knowledge requirements, while the commonly mentioned external sources were

---

[62] Shanahan, 2018b.

[63] HCI, 2018d.

industry rotations and custom-designed offerings from business schools. Other forms of T&D that played a role but were less often mentioned were service school offerings, courses and degree programs offered by colleges and vendors, and offerings from corporate university partners (e.g., Boeing's Leadership Center). Overall, the evidence suggests that DoD is utilizing a number of different external T&D providers and is engaged in efforts to incorporate industry expertise into internal T&D through the use of external training materials and presenters and the inclusion of industry participants.

Whether additional use of external T&D resources is needed to address gaps in Section 843–related knowledge is unclear. Interviewee perspectives suggest that many do see a need for additional use of external T&D resources. For industry rotations and the incorporation of industry resources into internal training, there was a general consensus around the value of increasing opportunities. Opinions were mixed for educational offerings from colleges and commercial vendors and professional certifications, with some arguing that additional opportunities would be valuable, while others perceived that current offerings were sufficient.

To fully determine whether additional external T&D is needed to address gaps in knowledge, we suggest that additional analysis is needed. A more robust needs analysis to determine what the exact knowledge requirements are in these areas and whether gaps do in fact exist and for whom is a first step, followed by a more explicit analysis of DoD's portfolio of internal and external offerings according to a common set of criteria. Findings from our interviews suggest that external T&D may contribute to DoD's portfolio in several ways: by increasing opportunities for applied and immersive learning and by ensuring that personnel are exposed to experts in industry. However, external T&D may be more challenging to tailor to the DoD context, may be more resource intensive, and may face challenges in terms of funding stability and legal and regulatory barriers. There are also significant capacity constraints to some of the external T&D offerings perceived to be most valuable. Conducting further analysis to make comparisons across internal and external T&D options can help to determine whether the value added offered by external T&D outweighs its costs.

CHAPTER FIVE

Approaches to Gauging the Effectiveness of External Training and Development

In Chapter Four, we laid out a range of criteria that DoD might use to determine whether external T&D should play a bigger role in addressing knowledge gaps in the areas of business acumen, industry operations, and industry motivation, and data on effectiveness are critical pieces of that decisionmaking. Accordingly, in this chapter, we discuss DoD's efforts to evaluate the effectiveness of its external T&D offerings. We describe some of the limitations to DoD's current efforts to formally evaluate external T&D and conclude the chapter with observations about possible ways to enhance DoD evaluation efforts.

An Overview of Common Practices for Evaluation of T&D

Before discussing DoD's in-use approaches for evaluation of its T&D, it is useful to account for the practices that are commonly used to evaluate T&D in organizations. We conducted a review of the literature on corporate T&D and the practices of six large organizations to identify common evaluation practices.

Evaluation of Impacts at Multiple Levels

According to the literature, evaluation is an important best practice for ensuring effective T&D,[1] and a diverse set of models or frameworks for evaluation of T&D are described therein.

The Kirkpatrick Model was the most commonly cited approach in the literature and was described by several resources as being the industry standard.[2] OPM used the Kirkpatrick Model as the framework for its *Training Evaluation Field Guide,* and several large-scale surveys of corporate T&D evaluation were structured around the

---

[1]   Wentworth et al. 2009; Gurdjian, Halbeisen, and Lane, 2014; Salas et al., 2012; Bernhard and Ingols, 1988.

[2]   Arthur et al., 2003; Executive Development Associates, undated; Wentworth et al., 2009.

Kirkpatrick Model.[3] The approach was developed in the 1950s and lays out four levels of evaluation, as follows:

1. Reaction (level 1): the degree to which participants find the training favorable, engaging, and relevant to their jobs
2. Learning (level 2): the degree to which participants acquire the intended knowledge, skills, attitude, confidence, and commitment based on their participation in the training
3. Behavior (level 3): the degree to which participants apply what they learned during training when they are back on the job
4. Results (level 4): the degree to which targeted outcomes occur as a result of the training and the support and accountability package.[4]

Other T&D evaluation models cited in the literature all emphasized the importance of measuring T&D at multiple levels, and many had level designations that corresponded to the Kirkpatrick Model in terms of measuring aspects of individual reactions, learning, and behavior and organizational results.[5] Some models, such as the context-input-reaction-outcome (CIRO) approach and the context-input-process-product (CIPP) approach, place a greater emphasis on looking at the context for the T&D and what goes into the implementation of T&D.[6] Other models emphasized the need to look at impacts beyond the individual and the organization, such as societal impacts.[7] Further, some approaches, such as Phillips' Five-Level Return on Investment (ROI) Framework, suggest that it is important to incorporate costs into evaluations of effectiveness to fully capture the returns on investment offered by various forms of T&D.[8] The Kirkpatrick Model is sometimes adapted to incorporate a fifth evaluation level for ROI, referred to as the Kirkpatrick/Phillips approach.[9]

According to a 2015 ATD survey, 96 percent of surveyed organizations evaluate the effectiveness of training in some way, though not all organizations were evaluating

---

[3]   Executive Development Associates, undated; OPM, 2011; Association for Talent Development, 2016b.

[4]   H. Topno, "Evaluation of Training and Development: An Analysis of Various Models," *Journal of Business and Management*, Vol. 5, No. 2, 2012, pp. 16–22; S. H. Lee and J. A. Pershing, "Evaluation of Corporate Training Programs: Perspectives and Issues for Further Research," *Performance Improvement Quarterly*, Vol. 13, No. 3, 2000, pp. 244–260; Kirkpatrick Partners, "The Official Site of the Kirkpatrick Model," undated.

[5]   Descriptions of models were retrieved from OPM, "Training and Development Policy Wiki," 2016; Topno, 2012; and Lee and Pershing, 2000.

[6]   Lee and Pershing, 2000.

[7]   Lee and Pershing, 2000.

[8]   J. J. Phillips, *In Action: Measuring Return on Investment (Vol 1)*, Alexandria, Va.: American Society for Training and Development, 1994.

[9]   See Association for Talent Development, 2016b, for an example of a survey using a Kirkpatrick/Phillips framework.

at multiple levels.[10] Organizations commonly evaluated reaction and learning, with more than 80 percent of organizations reporting evaluation at these levels. Evaluation at higher levels of the Kirkpatrick Model drops off significantly, with 60 percent reporting evaluation of behavior, 35 percent reporting evaluation of results, and 15 percent reporting evaluation of ROI.[11] According to a study that examined the evaluation practices of federal agencies, evaluation at higher levels was less common, but there were agencies (such as the IRS) engaged in evaluation of T&D results on measures of organizational effectiveness.[12]

While organizations less commonly measured behavior and results, they were more likely to see this information as valuable. A 2009 study found that 75 percent of organizations view evaluations of behavior and results as having high or very high value, compared with 59 percent viewing ROI as valuable, 55 percent viewing evaluation of learning as valuable, and 46 percent viewing reactions as valuable.[13] Evaluation at higher levels is more challenging for various reasons, including the cost of data collection, time lags before results materialize, and the challenge of tying behavior and organizational performance to specific T&D.[14]

Many Options for Evaluation Approaches and Measures

Within each Kirkpatrick Model level of evaluation, there is a wide range of approaches and measures that organizations use to assess T&D. Organizations must make decisions about the types of outcomes they want to measure (e.g., customer satisfaction, productivity) and the approaches to collecting data (e.g., survey, analysis of personnel data). A sample of the various approaches that were described for T&D evaluation is provided in Table 5.1.

There were relatively few options provided for evaluating effectiveness at Level 1. Organizations most commonly relied on end-of-course surveys, though more in-depth efforts to conduct focus groups or interviews were also described as options for assessing participant reactions. T&D instructors can also assess reactions in real time using quick checks for feedback during the training.

The literature described a broad range of approaches to assessing learning. These included formal assessments that could be administered to participants, such as in-course assessments and technical certification exams. More-qualitative approaches to assessing learning through case studies, observations, interviews, focus groups, and learner/participant presentations were also mentioned. Processes embedded into the

---

[10] Association for Talent Development, 2016b.

[11] Association for Talent Development, 2016b.

[12] Kirkpatrick and Kirkpatrick, 2012.

[13] Wentworth et al., 2009.

[14] Wentworth et al., 2009.

Table 5.1
Sample of Measures and Approaches to Evaluating T&D

Level of Evaluation	Sample Measures	Sample Data Collection Methods
Reaction (Level 1)	• Employee satisfaction • Employee engagement • Trainer impressions	• Surveys of participants • Pulse check • Participant interview • Participant focus group
Learning (Level 2)	• Increase in employee knowledge or intellectual capability • Level of Bloom's taxonomy learning • Employee engagement • Manager perceptions • Impact maps • Impressions of trainers • Return on expectations • Seat count and number of hours delivered • Technical certification	• Knowledge test • Skill observation • Presentation or teachback • Action planning, action learning • Case study • Benchmarking against local metrics or industry standards • Survey • Interviews • Focus groups
Behavior (Level 3)	• Work activities • Customer perceptions of behavior • Manager perceptions of behavior • Employee productivity • Performance ratings • Proficiency or competency levels	• Behavior observation • Work review • Request for supervisor feedback • Action planning • Performance records monitoring • Program follow-up session • Follow-up survey • Interviews • Focus groups • Customer/client assessments
Results (Level 4)	• Customer satisfaction • Employee engagement • Product or service quality • Time or resource savings • Customer feedback • Employee morale • Grievances, errors • Financial performance • Employee morale • Turnover, absenteeism • Perceptions of impact • Proficiency or competency levels • Productivity indicators (e.g., time, output per employee)	• Action planning • Work review • Request for validation (supervisor feedback) • Review of key business and human resources metrics • Peer evaluation • Follow-up surveys • Interviews • Focus groups

SOURCES: Resources used to create this table include OPM, 2011; Pearson, undated; Wentworth et al., 2009; C. Anderson, "Slowly, Steadily Measuring Impact," *Chief Learning Officer*, Vol. 12, No. 5, 2013, pp. 52–54; and OPM, *A Guide to Strategically Planning Training and Measuring Results*, Washington, D.C., 2000.

NOTE: We did not include ROI measures and methods in the table because they are complex to describe and all focus on a single measure, ROI.

design of T&D, such as action planning and action learning, were also described as approaches to capturing the learning that occurs from T&D. In addition to collecting data directly from participants, the literature suggested that supervisors or trainers could also provide valuable data for the evaluation of learning.

Surveys of participants were a common approach to measuring behavioral changes; a 2009 study found that approximately one in three organizations conducted them to evaluate impacts of T&D on behavior.[15] Many evaluation approaches for examining behavioral impacts (Level 3) were similar to those described for Level 1 and 2 evaluations, including observations, interviews, focus groups, employer surveys, and action planning. New strategies that were mentioned included review of work products and personnel records and the use of client and customer feedback.

The organizational (Level 4) impacts of interest are likely to vary by organization and may be somewhat different for private-sector companies versus government agencies. The literature suggested many possibilities, including client feedback, financial performance, employee turnover, and employee satisfaction. The most commonly reported measures on a 2009 survey were customer and employee satisfaction measures, learner perception of impact, proficiency or competency levels of employees, and supervisor perceptions of impact.[16] Productivity indicators, turnover, and actual business results were somewhat less common but were reported as being used by 22 to 26 percent of organizations to evaluate T&D effectiveness.[17]

Finally, there is a range of different approaches to calculating ROI. As noted previously, Phillips is credited with introducing an ROI approach that is commonly used to evaluate T&D.[18] OPM provided a basic approach to measuring ROI in *A Guide to Strategically Planning Training and Measuring Results*.[19]

DoD's Current Evaluation Practices

In many ways, DoD's current practices for evaluating its internal and external T&D mirror what we found in the broader review of literature and practices of large organizations. Two DoD interviewees reported use of the Kirkpatrick Model as the evaluation framework, so we used that framework for our discussion of evaluation practices. Interview evidence suggests that evaluation at Level 1 was most common and that evaluation was infrequently occurring at higher levels. Interviewees described evaluation of behavior being conducted for internal T&D, such as DAU courses, but we

---

[15] Wentworth et al., 2009.

[16] Wentworth et al., 2009.

[17] Wentworth et al., 2009.

[18] Phillips, 1994.

[19] OPM, 2011.

did not find evidence of this type of evaluation for external T&D. The most common approach to evaluating both reactions and behaviors was surveys of participants. We provide a more detailed description below of the various types of evaluation practices that were used to evaluate internal and external T&D.

Evaluation of Internal T&D

DAU makes an explicit effort to evaluate its T&D offerings at all four levels of the Kirkpatrick Model.

> The director of Strategic Planning and Learning Analytics is responsible for developing the strategic plan and annual performance plan for the university. The director is also responsible for end-of-course/event survey administration, data collection and analysis, and program evaluation of learning assets at each of the four levels in the Kirkpatrick model. (DAU Catalog, p. 14)

DAU also lays out goals for its evaluation efforts as follows:

> A continuous robust evaluation program for our courseware and curricula helps ensure content is current and relevant and teaching and learning methods are designed to produce expected learning outcomes.[20]

To assess course offerings across the various levels of the Kirkpatrick Model, DAU collects data from students, faculty, and student supervisors. At the end of each course, students are asked to complete a survey that assesses their reactions to the course. In addition to this end-of-course survey, students are also given the opportunity to complete follow-on surveys 60 days after completing the course that allow for further reflection on reactions, learning, and behavior. Learning is assessed within the courses through various types of assessments, and faculty reflect on the results of these assessments. For 300- and 400-level courses, 120 days after course completion, supervisors of students are asked to complete a survey assessing the impact of the instruction on a student's behavior.[21]

For the students, surveys include responses for levels of agreement with comments such as "This training has improved my job performance" and "This training will improve my job performance." For supervisors, surveys include responses for levels of agreement with comments such as "This training has improved the employee's job performance."[22] Interviewees said that this helps, for example, "find out, okay, are things going better on the job? Did we help you in some way be more efficient, more speed, more savings in what you're doing?" (DoD 9).

---

[20] DAU, DAU Directive 701, *Curricula and Program Evaluation*, January 14, 2013.

[21] DAU, DAU Directive 701, 2013, p. 4.

[22] These examples are from sample "Metrics that Matter" reports provided by DAU.

In addition to the evaluation of reactions through student surveys,[23] interviewees also described data collected on participation in continuous learning courses that are also used to gauge whether they are being used effectively by participants:

> We do lots of in-depth analytics that can tell how long people have lingered on a particular site or whether they jump from site to site within the DAU portfolio. Our workflow learning folks do track that kind of thing to [determine whether] this truly value added in that sense. (DoD 8)

Evaluation of External T&D

The evaluation of the impact of external training options can be difficult. For degree programs and executive education programs, interviewees indicated that student assessments were the primary source of information.

> For our [executive education course] I send an email. I require them to, at the end of the course, send me an email and tell me the things that they learned and how they're going to do their job differently because of the course. So I only have examples from individuals to say that it's effective. (DoD 2)

> Well, naturally from the staff level, we maintain those relationships and partnerships [with universities] to know what's going on. Every single participant has to produce a report and tell us the value of their efforts and what they were working on, as well as make a recommendation for if you had to do it all over again, would you send yourself or you know the next guy to this job? So there is a high degree of taking care of the next person. (DoD 5)

Those we interviewed had varying responses to the question of how the effectiveness of industry rotation programs is assessed. One industry representative indicated that "[t]here's not a lot of communication between [the corporation's] managers directly to the DoD or their career managers within the military" (Industry 10), while another said that for his company "there's definitely a formal, if you will, report at the end around the training . . . like basically a performance report that we send back" (Industry 11).

The latter is consistent with a comment about the Information Technology Exchange Program (ITEP) that "we have an evaluation or a summary that we ask for from the students and we also get it from the sponsor for them to come back and tell us how effective they thought the exchange program was for their organization" (DoD 22).

Unfortunately, evaluations have not been collected and analyzed in a systematic way, and neither the DACMs nor SMEs we interviewed in other DoD organizations had information on how they gauged the effectiveness of external programs.

---

[23] DAU, *DAU E-learning Asset Development Guide*, October 31, 2008.

Efforts to change this may be under way. A fairly new (2016) DoD Instruction, *Legislative Fellowships, Internships, Scholarships, Training-with-Industry (TWI), and Grants Provided to DoD or DoD Personnel for Education and Training*, mandates that for fellowships, internships, scholarships, training-with-industry programs, and grants for education and training, military departments

> [w]ork collectively to design, implement, and operate a standardized system for collecting, analyzing, and interpreting direct cost data and performance metrics to evaluate their programs and to prepare annual reports.[24]

The Office of the Under Secretary of Defense for Personnel and Readiness has begun collecting data from the services on participation in and costs related to these programs, but measures of performance and effectiveness remain elusive, and services have provided varying types and amounts of information to meet the instruction's requirements.

We found two attempts to quantify the effectiveness of rotational programs that could suggest approaches for future efforts to evaluate them. One was a 1989 AFIT thesis that surveyed officers who had participated in the Air Force's EWI program over a period of four years to compare several attitudes (such as the intent to remain in the Air Force) with those of individuals who had not participated in the program.[25] The other was a 2017 Naval Postgraduate School MBA Professional Report that proposed a systematic approach to a cost-benefit analysis of training with industry programs.[26] Both indicate the potential importance of tracking individuals in their post-training careers in order to assess the value of the external training experience.

The Effectiveness of External T&D

As described above, DoD efforts to evaluate the effectiveness of participation in external T&D were limited and uneven across different types of training. The lack of effectiveness data prevented us from being able to conclusively speak to the effectiveness of external T&D. In this section, we first lay out these limitations, followed by a descrip-

---

[24] DoD, DoDI 1322.06, *Fellowships, Legislative Fellowships, Internships, Scholarships, Training-with-Industry (TWI), and Grants Provided to DoD or DoD Personnel for Education and Training*, USD/P&R, October 12, 2016b.

[25] E. R. Hernandez, *A Study of Benefits Resulting from the AFIT Education with Industry Program*, AFIT thesis AFIT/GSM/LSR/89S-18, September 1989.

[26] M. S. Flynn and A. Souksavatdy, *Return on Investment for the United States Navy's Training with Industry Program*, Naval Postgraduate School MBA Professional Report, June 2017. The approach has many of the characteristics outlined in OPM's *A Guide to Strategically Planning Training and Measuring Results*.

tion of the qualitative evidence on the effectiveness of external T&D that was gathered through this study.

Current Limitations to Fully Evaluating Effectiveness of AWF T&D

Numerous challenges made a formal assessment of effectiveness of external T&D difficult to conduct. We identified these challenges through a review of literature on T&D evaluation and interviews with DoD and industry leaders. In many cases, these challenges were not specific to DoD or to external T&D and simply reflect the difficulty of evaluating T&D for all organizations, including those in the private sector. We summarize the key challenges below.

There were limited needs analyses available to inform evaluation. To evaluate whether T&D is effectively meeting needs, a needs analysis is an important first step because it determines what knowledge needs to be built and what goals DoD intends to accomplish with T&D.[27] The Section 843 concepts of business acumen, knowledge of industry operations, and knowledge of industry motivation were proffered by Congress, and, as we noted earlier in this report, they have not been clearly defined within DoD. In Chapter Two, we relied on the inclusion of related terminology in existing competency models to identify evidence of possible need for these particular types of knowledge, but the career fields have not systematically assessed who needed this type of knowledge and at what level of proficiency. If the career fields fully identified the need for these specific types of knowledge and incorporated requirements across competency models where necessary, then our analysis could better capture knowledge requirements in these areas.

Participation in various forms of external T&D is not tracked in personnel files systematically. We discussed limits to tracking of participation in T&D in Chapter Four. Some effectiveness evaluations may require the use of administrative data to examine career outcomes, making it important to be able to identify the range of T&D experiences an individual has had and pinpoint the populations of personnel reached by various forms of T&D, including both internal and external T&D participation. Personnel files that identified certifications and degrees held and other forms of participation in T&D would have been helpful in detailing the reach of T&D and assessing the effectiveness of T&D through analysis of personnel data.

The measures used to assess effectiveness are limited and largely focused on reactions and learning. According to interviewees, evaluations of external T&D for personnel focused on student surveys and presentations that primarily documented the participants' perceptions about their training. External T&D providers and DoD leadership occasionally described assessments or anecdotal feedback they received on organization impact of a training activity, but these data were not being used to formally assess the effectiveness of training. As described in the literature on the Kirk-

---

[27] Salas et al., 2012.

patrick Model, these measures of impact on "reactions" and "learning" are the most basic measures and are not sufficient to fully understand whether T&D is effectively meeting DoD needs.

Isolating the impacts of specific T&D activities on behavior and organizational effectiveness is challenging. Individual behavior changes and organizational changes are more challenging to link to specific T&D experiences because T&D is one of many factors that may be playing a role in shaping work behaviors and organizational performance. Individuals bring their own personal experiences, preferences, and constraints into the workplace and also face a range of external factors that may shape behavior and organizational effectiveness, such as managerial support, collaborative work of team members, client behaviors, and legal requirements. According to interviewees, it was difficult to account for all of these other important factors (without rigorous analytic techniques). Measures of reaction and learning may be easier to pinpoint because they are typically assessed close to the time of T&D and tightly linked to the T&D experience.

The time lag before behavioral and organizational results materialize makes evaluation challenging. Reactions of participants and learning of knowledge could be measured during or immediately after T&D activities, while other results at the behavioral and organizational levels may have taken years to materialize. And decisionmakers often prefer data to determine whether T&D investments are paying off on a shorter timeline. In addition, following T&D participants and collecting data over the longer timespan is likely to be costlier. Time lags in results also make it hard to isolate the impacts of specific T&D activities because many other things happen in the period between T&D participation and when results are observed.

Resources for evaluation are limited. As is common in many organizations, the resources devoted to evaluation of AWF T&D appear to be limited. When DoD worked with external organizations to deliver T&D, these programs were often reported to be overseen by a single individual, and resources may not have been set aside for the regular oversight and evaluation of these external T&D programs. In addition, there were no centralized resources for evaluating the value of earning degrees and professional certificates among DoD acquisition personnel.

Interview Perspectives on the Effectiveness of External T&D

While we were unable to fully assess the effectiveness of external T&D, either to confer Section 843–related knowledge in general or to close specific knowledge gaps. comments made by our DoD and industry interviewees regarding the usefulness and value of external T&D shed some light on perceived effectiveness. We first describe findings regarding the types of knowledge that external T&D opportunities were perceived to be effective in building, followed by findings on perceptions of effectiveness by type of external T&D.

Overall, evidence from interviews suggests that external T&D resources were perceived as valuable in building business acumen, knowledge of industry operations, and knowledge of industry motivation. First, interviewees described external training opportunities as providing DoD acquisition personnel with an opportunity to be "on the other side of the table" (Industry 17) to better "understand those thought processes in how industry views us" (DoD 8). According to interviewees, this alternative perspective enhanced the ability of AWF personnel to elicit better deals and performance from contractors. Interviewees also described external T&D as being important for deep learning about industry operations. Business school programs at colleges and universities were described by some interviewees as being an optimal place for building this type of knowledge. For example, "If you want to learn about how companies operate from a business perspective, you're going to get that at a business school" (Industry 14). Finally, interviewees mentioned that external T&D offered the opportunity to learn about industry best practices in the areas of business processes, leadership, and technology. "You have that cross-pollination, the best practices from industry, bringing them over to DoD" (DoD 7).

We examined a number of different types and providers of external T&D and perspectives regarding the value of these external resources varied. We provide a sample of interviewee perspectives by training type in Table 5.2, and we highlight some of the key findings by type of training below.

Industry Rotations/Exchange Programs

Our interviewees extensively discussed the role of industry rotations in conferring business acumen, knowledge of industry operations, and knowledge of industry motivation. While many courses and degree programs aimed to build in experiential components that provided hands-on experience with industry, some interviewees suggested that there was no substitute for being embedded in industry and engaging in work over a sustained period of six months or more. Interviewees reported that rotations with defense contractors and non-defense contractors were valuable and suggested there were efforts under way to expand the options for industry rotations, especially for civilians. There were some interviewees, however, who suggested that the value of industry rotations might be quite variable and raised concerns that these opportunities did not consistently provide returns beyond the individual and were not necessarily scalable.

Courses and Degree Programs from Universities and Colleges

Offerings from colleges and universities were also cited as a helpful resource for building business acumen, knowledge of industry operations, and knowledge of industry motivation. Interviewees discussed educational offerings interchangeably, making it difficult to distinguish perceived differences in terms of value for courses, executive education programs, and degree programs. Interviewees perceived these programs as complementary alongside DAU offerings, providing an opportunity for a deeper level of knowledge than might be provided by internal courses. In addition, interview-

Table 5.2
Interviewee Perspectives on Benefits of Using External Resources for T&D

Type of External Resource	Interviewee Quotes
Industry rotations and exchange programs	"It's in [Education with Industry] that people learn that the craft of industry and how industry operates and what motivates industry. So for folks inside of the government, you've got to either give them exchange opportunities and experiences to go out there and learn side by side with the folks in industry so that they can see what it's about." (Industry 7)
	"Half of the colonels in [military service] contracting were products of the Education of Industry Program. . . . And all of them, if you talk to them about it, would consider their time with Industry as a highlight of their career." (Industry 18)
Executive education and degree programs	"University of North Carolina is not defense industry but they have some pretty good courses on management and technology and culture and factors that are considered through a number of different great guest speakers that come in and talk about considerations in industry and interrelationship between industry and government." (DoD 21)
	"If you want to learn about how companies operate from a business perspective, you're going to get that at a business school, not from DAU. At least what you'll get from DAU, I don't think would be as substantial. I think it's scratching the surface." (Industry 14)
	"I'll use Logistics as an example; you have universities like Penn State, which is really good in the area of supply chain management. They're not competing with us. They're complementing us in terms of capabilities or expertise." (DoD 8)
Professional certifications and commercial training	"I have not seen a whole lot of great return on short-term professional training. Such as a . . . go get your project management professional certification by taking our course. I have not seen a whole lot of positive return on that." (DoD 10)
	"I believe leveraging industry and/or at least working closely with industry training providers, et cetera, to deal with, agree on, this consistent industry-wide standard and set of training standards would allow the DoD to utilize its resources a little bit differently and focus on the unique stuff while utilizing sort of this common base of industry stuff to do sort of what's really foundational and core learning." (Industry 7)
	"I would say it [adding more external T&D] is a balance. Potentially a little more, I think there is a great position for commercial training and commercial certification. I strongly disagree with those that would say just outsource it all." (DoD 9)

SOURCE: 2018 RAND Section 843 study interviews.

ees highlighted the perception that there was greater Section 843–related knowledge expertise among individuals teaching at colleges and universities. Finally, the opportunity to sit side by side with industry participants in these courses was perceived as important. However, others were not convinced that the offerings from colleges and universities were valuable. These interviewees suggested that DAU could likely provide equivalents to some of the courses and executive education programs offered by external providers and argued that internal offerings might be more valuable because they could be tailored to DoD's unique context at lower cost.

Professional Certifications

Perspectives on the value of professional certifications for developing business acumen, knowledge of industry operations, and knowledge of industry motivation were mixed. Professional certifications did not offer the immersive, experiential learning opportunities that many saw as the value added of external T&D. On the other hand, some interviewees believed that it was valuable for personnel to pursue professional certifications alongside DAWIA certifications as a form of continuing education, and some services and career fields actively encourage personnel to obtain professional certifications. According to some interviewees, these professional certifications could ensure that DoD personnel have mastered the knowledge that industry has determined to be the standard for the profession. Some interviewees argued that DoD might take a closer look to see how professional certifications and DAWIA certifications could be better integrated for career development.

Future Efforts to Assess the Effectiveness of External T&D

While we were unable to rigorously assess the effectiveness of external T&D for the purposes of this study, there are several things that DoD could do to strengthen evaluation efforts going forward. We briefly touch on these suggestions here and revisit them in the recommendations section.

We suggest four ways in which evaluation efforts could be enhanced to better understand the effectiveness of external T&D:

1. **Conduct a full needs assessment.** While previous chapters noted the value of a full needs assessment for the purposes of developing and implementing T&D, a needs assessment is also a critical input for an evaluation plan. Understanding the needs that T&D is addressing is helpful in identifying measures to track effectiveness in meeting those needs.
2. **Track all T&D participation.** If all participation in internal and external T&D were tracked at the individual level in administrative files, these data could be used in a range of ways to assess participation in and effectiveness of T&D.
3. **Apply internal evaluation policies for external training.** According to interviewees, DAU routinely evaluates reactions and behavior through participant and supervisor surveys. These evaluation processes could be adopted at relatively low cost and adapted to evaluate a range of custom and general T&D opportunities internally and externally. This would allow for comparable data across internal and external T&D and follow-up on behavioral changes for external T&D participants that is not currently happening.

4. **Consider more-rigorous evaluations for costly programs.** Interviewees noted that rotations and custom-designed executive education programs were costly, yet there was limited assessment of these T&D offerings. It may be particularly important for DoD to assess the value of the largest and most expensive programs to ensure sufficient ROI on these offerings.

Summary

Information on the effectiveness of all T&D offerings is needed to determine whether and how these resources should be used to build business acumen, knowledge of industry operations, and knowledge of industry motivation within the AWF. While evidence suggests that DoD aims to evaluate the effectiveness of T&D at multiple levels of the Kirkpatrick Model and conducts regular surveys of participants and supervisors to assess reactions and behaviors related to DAU courses, overall, we found that its efforts to evaluate external T&D were inconsistent and limited. More could be done to understand both who is participating in these opportunities and how participation impacts them.

While we did not have rigorous and consistent objective measures of external T&D effectiveness, interviewees perceived external T&D as being effective in building knowledge in all three types of Section 843–related knowledge. Perceptions of effectiveness varied by type of external T&D, and interviewees were most likely to emphasize industry rotations as being effective in conferring business acumen and knowledge of industry. In addition, while many cited university and college offerings as being effective, some argued that DAU could offer similarly effective T&D. Perspectives on the effectiveness of professional certifications were mixed. Some viewed certifications as valuable, while others argued that they did not offer the experiential elements that were perceived as important elements of the industry rotation and executive education offerings. However, these perceptions are not sufficient to determine whether additional external T&D could be effective in filling gaps related to Section 843. A full cost-effectiveness analysis of DoD's internal and external T&D portfolio, as described in Chapter Four, would be the most rigorous way to address the question of effectiveness.

Limited data on effectiveness of T&D is a common problem across many organizations, and the literature describes a number of challenges to evaluation of T&D. Our interviewees cited many of these challenges in evaluating DoD's T&D offerings. Yet, despite these limitations, there are some basic actions that DoD could take to enhance its capabilities to assess effectiveness, such as tracking participation at the individual level in personnel files and applying internal evaluation practices to external T&D. With well-defined goals for T&D, some routine data collection, and targeted evaluation of the more expensive external T&D offerings, DoD could substantially improve its ability to assess the value of external T&D.

CHAPTER SIX

Conclusions and Recommendations

In FY 2017 NDAA Section 843, Congress directed USD(A&S) to assess business-related training for the AWF by examining current sources of training and career development opportunities related to business acumen, knowledge of industry operations, and knowledge of industry motivation, gaps in these three types of knowledge, and the role that non-DoD (i.e., external) organizations could play in addressing these gaps.

RAND researchers used multiple approaches to accomplish these tasks, including a review of AWF competency models, interviews and targeted discussions with SMEs, a literature review of relevant studies and guidance, reviews of documents and websites related to DoD and non-DoD sources of training, and a review of approaches used by DoD and non-DoD organizations to determine training effectiveness. The tight schedule required to meet the congressional deadline made it infeasible to gather information from individual AWF personnel and limited the number of expert interviews and follow-up interviews we could conduct.

Conclusions

We had difficulties in assessing gaps in Section 843–related knowledge that we did not foresee at the outset of our study. Specifically, these terms are not formally defined, nor are they consistently included in the competency models that drive the creation of DAU learning assets and AWF T&D strategies more generally. The lack of a documented need and desired proficiency level for each type of knowledge rendered it challenging to assess the extent to which they were lacking in the AWF workforce overall, much less on a career-field or position basis. Moreover, few efforts to systematically assess competency gaps within the AWF have been conducted in recent years, and the competency assessments that HCI and DCPAS have planned using the DCAT were just getting under way at the time of this study. Accordingly, we relied on expert interviews with DoD leaders and external stakeholders to identify perceived gaps in business acumen, knowledge of industry operations, and knowledge of industry motivation and sought to corroborate interview findings with previous gap-related studies as much as possible.

Section 843–Related Knowledge Gaps Exist, But Their Extent Is Unclear

Based on those interviews and the studies we reviewed, we conclude that gaps related to business acumen, knowledge of industry operations, and knowledge of industry motivation are present within the AWF to an indeterminate extent. Specific aspects of business acumen cited in interviews include risk management and earned value management, and aspects of industry operations–related knowledge perceived as both important and lacking include financial practices, supply chain management, small business, agile development, and cybersecurity. The gaps in industry-related knowledge were perceived as having an influence on other types of knowledge, skills, and abilities important to the AWF: negotiation, developing and understanding requirements, and cost and price analysis.

Without Efforts to Better Estimate Needs and Effectiveness, Gaps May Persist

In assessing knowledge gaps and determining whether external T&D might be used to address gaps in training, we encountered challenges that we documented in earlier chapters. In fact, many of these challenges also potentially contribute to the knowledge gaps we identified.

Lack of Defined Knowledge Requirements in Section 843 Areas

Our analysis suggests several possible reasons for these gaps. First, in Chapters One and Two, we noted the lack of formal, widely used definitions for *business acumen*, *industry operations*, and *industry motivation*, and only the first type of knowledge, business acumen, is expressly referred to in a small number of career field–level competency models. It is unclear whether these knowledge types are not included in these models because the career fields are not perceived as needing them or because of an oversight of some sort (e.g., the belief that they are covered in other requirements). We also considered the possibility the Section 843 knowledge types are covered using different terminology. Accordingly, we developed and operationalized definitions for them; used keywords based on those definitions to analyze the models more deeply for evidence of Section 843–related knowledge requirements; and found that certain aspects, such as human capital management and negotiation, are frequently included in the models. This suggests that the Section 843 knowledge requirements are at least partially determined, but, again, our keywords may not have fully covered what the knowledge types encompass. An additional concern is that most of the competency models do not include a required proficiency level for the competency elements included therein. This means that even if business acumen was included in a model, for example, it is not specified whether professionals in that career field need to be proficient at level 3 (intermediate) or level 5 (expert). Thus, a true "yardstick" against which to gauge AWF actual proficiency levels in business acumen, knowledge of industry operations, and knowledge of industry motivation is lacking.

In addition, the lack of clearly defined knowledge requirements leads to challenges in determining whether the number of current external T&D offerings is sufficient. While we found that DoD is using a range of external T&D offerings, it is possible that some T&D opportunities are currently underutilized: For example, the number of DoD personnel participating in industry-based rotational assignments is low relative to the size of the AWF. But we could not determine what the right number should be without well-defined knowledge requirements that define who needs Section 843–related knowledge.

Lack of Formal, Standard Gap Assessments on Section 843–Related Knowledge

In Chapter Three, we identified another possible reason for these gaps: a limited number of efforts by DoD to assess proficiency related to Section 843–related knowledge. Using multiple search strategies, including asking interviewees for studies they were aware of, we were only able to locate one recent career field–level study, a Program Manager–focused study conducted in 2014, and two unpublished service-level studies. It is possible that we did not query all of the knowledgeable parties in DoD during our short study time frame, but we did broach this topic with key leaders tasked with AWF T&D. A failure to assess gaps on a regular basis in a way that can be rolled up to the full AWF level means that they are neither detected nor addressed, and, left unchecked in this way, they may persist or even grow.

Limited Tracking of Participation in External T&D

We covered T&D activities in Chapter Four, and that chapter summarizes other factors potentially contributing to gaps in business acumen, knowledge of industry operations, and knowledge of industry motivation. First, DoD does not have a high level of awareness of who has completed T&D activities that are perceived to confer Section 843–related knowledge. For example, although business degree programs are regarded as a strong way to build these types of knowledge, DoD centralized personnel databases, such as DCPDS, are missing data on the types of graduate degrees held by AWF members. In another example, industry rotations and fellowships are tracked in different ways by the military services, and to date they have been unable to comply with the reporting mandates specified in DoDI 1322.06, *Fellowships, Legislative Fellowships, Internships, Scholarships, Training-with-Industry (TWI), and Grants Provided to DoD or DoD Personnel for Education and Training.* Finally, through interviews, we learned that individual personnel records do not typically include all relevant T&D activities. This lack of clarity on who has completed what forms of T&D may make it difficult for those tasked with AWF career management to identify training gaps and direct the people who most need training related to Section 843 knowledge to T&D opportunities.

Limited Efforts to Assess T&D Offerings as a Portfolio

DoD is making use of a wide array—and possibly the full array—of available T&D options, so a high-level training gap was not apparent (i.e., DoD is not completely over-looking a type of T&D used by industry). However, it is not clear that external T&D offerings were developed through an assessment of the current T&D offered to address particular knowledge requirements and an identification of specific gaps in current T&D offerings. This may lead to the duplication of offerings, the use of external T&D where internal T&D could have better met needs (or vice versa), and the failure to fill the most pressing gaps in current T&D offerings.

Our interviewees extensively discussed the value of external T&D activities (e.g., industry rotations, executive education, and degree programs offered by non-DoD colleges and universities), as well as the belief that external T&D should be used more extensively to build business acumen, knowledge of industry operations, and knowledge of industry motivation within the AWF. However, a determination of whether more external T&D is needed requires an accounting for the costs of this T&D and how the costs and benefits of various external T&D options compare to and complement internal T&D offerings. We did not receive any cost-benefit analyses of this nature, nor were such analyses referred to in our interviews. Analyzing the portfolio of T&D offerings in a more strategic and coordinated manner would allow DoD to ensure that external T&D is being used where it can provide the most value added.

Limited Data on the Effectiveness of T&D

Finally, a lack of T&D evaluations affects our—and DoD's—ability to determine whether current T&D offerings are effectively addressing knowledge requirements, which, in turn, affects our ability to assess knowledge gaps *and* training gaps. Effectiveness data are critical to the comprehensive analysis of DoD's T&D portfolio described previously because effectiveness data provide the ultimate measure of the value that each type of T&D offering brings to the portfolio. Having comparable data on effectiveness across different T&D offerings would be particularly useful for portfolio management. However, as we discussed in Chapter Five, in-use evaluation practices are relatively limited and focus primarily on participant reactions. For example, efforts to gauge the effectiveness of industry rotations were highly variable across fellowship programs, and, in general, evaluations that were conducted were based on the student or participant's own perception of the T&D's usefulness, how it affected his or her proficiency level, and how it affected his or her work performance. On the other hand, DAU is conducting more-robust analysis for a limited set of its own offerings: For DAU 300- and 400-level courses, the university surveys supervisors of course graduates 120 days after course completion to gauge how the course may have influenced the course graduates' work performance. These efforts remain sparse, however, and the limited evidence of effectiveness makes it difficult to determine whether the "right" mix of T&D is in use and whether desired knowledge gains are being attained.

DoD Is Attempting to Address Gaps

In spite of these barriers to measuring the extent to which Section 843–related knowledge gaps exist, DoD has undertaken efforts to avoid and close them. As we described in Chapter Four, ACQ 315, "Understanding Industry," was developed in response to training and knowledge gaps identified in two Program Management career-field studies. Other DAU courses and learning assets (e.g., workshops) have been created in response to perceived knowledge gaps as well. For example, DAU has added courses on agile development and software cost estimating to its curriculum and has traveled across the country offering workshops related to cybersecurity. External T&D resources have been deliberately tapped to close knowledge gaps as well, but seemingly to a very limited degree. A prominent example of this type of gap-related corrective action is the "Better Business Deals" course that the University of Tennessee was contracted by the Air Force to develop to address a perceived lack of knowledge of industry. Overall, efforts to close gaps tend to be course-based in nature and have a relatively limited capacity.

These findings, along with insights related to promising T&D practices and interviewee opinions of facilitators of and barriers to greater use of external T&D, informed the development of the recommendations presented in the remainder of this chapter. We have grouped our recommendations into three categories: process-focused recommendations for DoD, external T&D–focused recommendations for DoD, and recommendations for Congress.

Process-Focused Recommendations for DoD

Clarify the Nature and Extent of Needs for Business Acumen, Knowledge of Industry Operations, and Knowledge of Industry Motivation

Our efforts to identify the extent to which AWF personnel lack business acumen and industry-related knowledge revealed process-related obstacles to making this determination with precision. The career-field competency models and DAU courses required for DAWIA certification that we reviewed rarely used these exact terms, nor did the small number of knowledge gap–related studies that we located. Thus, DoD should clarify the nature and extent of needs for Section 843–related knowledge for each career field as an important first step to determining learning objectives and the correct mix of T&D.

While there is an OPM definition of business acumen, career-field FIPTs need to determine whether it comports with the AWF understanding of the term. For industry operations and industry motivation, ASD(A&S) should engage Congress to arrive at a shared understanding of what the terms mean. Armed with this information, FIPT members can then discuss in their annual updates how the terms map to existing competencies and whether new competencies need to be defined and incorporated. They can also use these definitions to help identify the degree to which members of each career field need this type of knowledge to perform their job tasks.[1] This is an important first step to determine whether a more-targeted focus on the areas identified in Section 843 is necessary or whether an explicit effort to build new competencies and assess gaps is required.

If a phased approach to determining needs more precisely across the career fields is needed, because of resource availability, DoD should consider prioritizing the career fields with a higher perceived need for Section 843–related knowledge, as suggested by our findings in Chapter Two. These findings, based on analyses of existing competency models and courses required for DAWIA certification, as well as our interviews, consistently suggested that Program Management and Contracting have a high relative need for these types of knowledge. Engineering and Life Cycle Logistics also may merit prioritization, based on our findings.

Improve Approaches to Competency Assessments and Models

Functional leaders, perhaps through the WMG, should coordinate the development of a standard format for competency models that conforms to the structure defined in DoDI 1400.25, Vol. 25, and DoDI 5000.66 and includes proficiency standards in line with the requirements in those instructions. Materials we received from Contracting career-field representatives may serve as a useful starting point because the competency model for that career field uses the five-level proficiency hierarchy for many of the competencies included therein. The format should include professional competencies that were excluded from AWQI, such as communication and the ability to work effectively with industry, but that can overlap with the types of knowledge specified in Section 843. Developing proficiency standards for both technical and professional competencies would facilitate the identification of knowledge needs and gaps.

The currently codified practice of conducting in-depth competency assessments every five years seems appropriate, as do the relatively informal annual reviews in accordance with DoDI 5000.66 that verify the adequacy of existing competencies. However, functional leaders should ensure that such assessments happen at the mandated intervals and should also consider developing criteria that could trigger in-depth assessments earlier than the five-year point. For example, the ongoing DCAT assessment could illuminate the need for competencies that are not yet included in FIPT competency models.

With Congress's interest in business-related training, functional leaders should consider additional ways to incorporate industry standards and perspectives into these models. For some career fields, there may be an industry certification that could inform competency model development or even become a key part of the model (e.g.,

---

[1] While this study focuses on knowledge gaps, we recognize that, in a practical sense, both knowledge itself and the ability to apply that knowledge are essential for successful outcomes.

as noted in Chapter Two, APICS and the Society for Logistics Management for the Lifecyle Logistics career field). Including representatives from industry on SME panels or as "associate" FIPT members could be another way to ensure that knowledge needs related to business acumen, industry operations, and industry motivation are sufficiently incorporated into models. Because needs for Section 843–related knowledge and the availability of relevant industry-based certifications vary among AWF career fields, the best way to infuse competency models with industry perspective will differ across them.

Improve Approach to Knowledge Gap Assessments

After DoD develops requirements for Section 843–related knowledge, including proficiency levels, determining how many AWF personnel meet those standards—and how far short others fall—is an important next step. Accordingly, DoD needs to develop and implement a rigorous approach to measuring proficiency in these areas—ideally, one that goes beyond relying on AWF professionals' self-reported proficiency, as was the case in the unpublished DoD studies that we reviewed as part of our analysis of Section 843–related knowledge gaps. For example, perspectives from supervisors and SMEs could also be included in the gap assessment. In addition, it may be possible to use proficiency tests to a limited degree, such as having personnel with a very high need for specific aspects of business acumen or knowledge of industry to participate in "Section 843 knowledge simulations" or scenario-based testing to gauge not only their level of knowledge but also their ability to apply it. Finally, to the greatest extent feasible, DoD should use standard methods and measures across career fields so that results can be aggregated to the full AWF level. This, in turn, would help to calculate the required capacity for gap-closing T&D activities. As with the needs determinations themselves, to the extent that sequencing is needed, in light of resource constraints, DoD should consider prioritizing for gap assessments those career fields with comparatively high needs for Section 843–related knowledge.

Improve Coordination of Internal and External T&D as a Single Portfolio of Offerings

While DAU's website has very good information about internal (DAU course and DAU-equivalent course) options, information about external T&D options is scattered among various DoD websites, DACM-maintained ones most notably. As Chief Learning Officer of the DoD acquisition community, the DAU president should consider maintaining a T&D resource directory that lists all training options (academic degrees, executive training programs, fellowship programs) available through DoD, the services, commands, and agencies. With better understanding of various options, AWF career managers and personnel supervisors can make informed decisions about what training resources are appropriate for the career plans of those they supervise, and the directory may empower individual professionals to do more to chart their own career development course.

While a centralized, DoD enterprise–level repository for T&D opportunities will help managers and supervisors to improve decisionmaking about which types of T&D are used at a micro level, there may also be a role for more centralized coordination and management of DoD's full portfolio of external and internal T&D offerings. It is important to ensure that DoD is utilizing external T&D in a strategic way to fill gaps in internal offerings rather than developing programs in silos that may be duplicative or may not address critical knowledge requirements. Without a coordinating body that is specifically charged with approving and overseeing external T&D offerings as part of the larger portfolio of T&D, these programs will continue to be adopted as one-off programs that may or may not be the most cost-effective way for addressing a particular knowledge requirement. It may not be the case that the day-to-day management of these programs is best done centrally, but the use of a more coordinated and intentional process for the identification of training gaps, the approval of new offerings, and the evaluation of offerings could help to ensure that DoD's overall portfolio of T&D is meeting training needs in an effective and cost-efficient way.

Improve Tracking of Participation in T&D Activities That Confer Business Acumen, Knowledge of Industry Operations, and Knowledge of Industry Motivation

People who join the AWF, particularly those at the mid-career level or higher, may already possess business degrees or experience working with industry. The ability to track this information in DCPDS for civilian personnel, who constitute the bulk of the AWF, would make it easier to determine which AWF personnel already possess some level of business acumen, knowledge of industry operations, and knowledge of industry motivation and which do not, which, in turn, may help to ensure that the personnel with the greatest need for more Section 843–related knowledge have priority to participate in capacity-limited T&D activities. Tracking personnel who complete relevant T&D while in the AWF would be useful for the same reason and also may facilitate evaluation of DoD's T&D offerings. Improvements to tracking likely will require information technology investments, such as an update in existing data systems or better integration of service-specific data repositories. Improvements in tracking also might require incentives for AWF personnel to provide and possibly enter relevant education and experiences into their records.

Improve Evaluation of T&D

A full needs assessment is necessary for developing and implementing T&D, but it is also a critical input for an evaluation plan. Understanding the needs that T&D is addressing is helpful in identifying measures to track effectiveness in meeting those needs. With this understanding, those tasked with AWF career T&D responsibilities (e.g., functional leaders, DACMs, and HCI leadership) should:

- **Track all T&D participation.** As suggested in the preceding recommendation, if all participation in internal and external T&D activities were tracked at the individual level in personnel data systems, these data could be used in a range of ways to assess participation in and effectiveness of T&D.
- **Apply internal evaluation practices to external T&D.** For a limited set of courses, DAU evaluates reactions and behavior through participant and supervisor surveys at different time intervals. These evaluation processes could be adopted at relatively low cost and adapted to evaluate a range of custom and general T&D opportunities internally and externally. This would allow for comparable data across internal and external T&D and follow-up on behavioral changes for external T&D participants that is not currently happening.
- **Track career outcomes of T&D participants.** Better tracking of participation in specific T&D activities would enable DoD to compare career outcomes, such as performance ratings, promotions, and retention, of those who have participated in Section 843 knowledge–related T&D with those who have not. This could be a useful complement to perception-based assessments of T&D effectiveness, particularly when efforts to document the impact of T&D on mission outcomes would be challenging, as is often the case.
- **Consider more-rigorous evaluations for costly programs.** Rotations and custom-designed executive education programs are costly, yet there is limited assessment of these T&D offerings. It may be particularly important that DoD assess the value of the largest and most expensive programs to ensure sufficient ROI on these offerings.

External T&D-Focused Recommendations for DoD

Clarify and Enforce Reporting Requirements for Fellowships and Industry Rotations

DoDI 1322.06 directs military departments to provide an annual report to USD(P&R) on their fellowships, internships, scholarships, training-with-industry programs, and grants to determine how the cost-effectiveness of each of these T&D options compares against others. Although the DoDI provides a standard reporting template that includes measures such as participant names, direct costs, and indirect costs, as of the completion of this study, the departments were not consistently using it; they either did not provide all of the requested information or it was unclear whether information, such as costs, was calculated in similar ways by the different departments. In addition, the guidance does not specify how a critical part of a cost-effectiveness analysis—T&D effectiveness in conferring knowledge and producing desired outcomes—should be determined and reported. Thus, we recommend that USD(P&R) develop guidance for estimating the ROI of these programs, to include specifying how indirect and direct costs should be calculated and the standards and methods to use in evaluating the

T&D upon which the DoDI focuses. Through our targeted discussions, we learned that an effort is under way to promote greater standardization of department inputs, but the scope and timeline of that endeavor is unclear.

Further Assess the Need for Government-Industry Co-Education

Interviewees commented on the need for more government-industry co-education in two areas: industry rotations and use of industry resources (i.e., participants, presenters, standards) in internal T&D. Adopting the process recommendations described previously will provide DoD with the tools it needs to more conclusively determine whether there are training gaps that require more of these industry T&D resources.

If there are gaps in government-industry co-education, ongoing efforts to enhance opportunities in these areas may help fill some of them. For example, the Public-Private Talent Exchange authority is developing new industry rotation opportunities. Recently, the Deputy Secretary of Defense encouraged "frequent, fair, even and transparent dialogue" with industry.[2] As reported in Chapter Four, AWF T&D is contributing to this effort in several ways, including the participation of some DAU instructors and industry rotation participants in in-house training offered by industry, such as the Boeing Leadership Center and the Lockheed Martin Leadership School; the invitation of industry participants to participate in DAU training; the use of guest speakers in DAU courses and other DoD T&D activities; and the incorporation of industry resources and standards to shape internal T&D content.

While we acknowledge that increases in co-education may be difficult to achieve and may require incentivizing industry cooperation in some way, there are some possible actions under DoD's purview to expand these opportunities. Companies may be willing to take on a small number of participants for rotations or participation in a company's in-house training, but it is possible that additional defense companies and companies outside of the defense industry would offer valuable T&D experiences for AWF personnel. An increase in the number of industry students in DAU in-residence classes could also be valuable; we noted that a relatively small number of industry professionals, 181 students, participated in DAU classroom courses in FY 2017. Many courses had little or no industry participation. which could be because the perceived benefit of specific courses to industry participants is low, industry willingness to cover the costs of participation is low, or course capacity is low and must be devoted to AWF members. However, some advanced courses have relatively high participation by industry already and suggest that this approach holds promise. For example, ACQ 404, "Senior Acquisition Management," a four-and-a-half-day in-residence course, had 11 industry participants out of a total of 68 graduates (16 percent) in FY 2017. It is also possible that lower-level courses with greater capacity may both appeal to industry and provide opportunities for government-industry interaction at an earlier career stage,

---

[2] Shanahan, 2018a.

thereby planting the seed for continued interactions through students' careers. Given the costs that both industry and government incur related to industry participation in DAU in-residence courses (e.g., missed time from a student's regular work, financial cost for the seat in a class), DoD will need to consider carefully which courses offer the greatest return on this type of co-education investment.

DAU already values industry experience in individuals it considers for faculty positions, and we recommend continuing or even expanding this practice (e.g., experience working in industry might be a requirement for more DAU positions). The use of guest instructors and guest speakers from industry is also a useful practice that should persist and potentially be expanded so that more DAU students are exposed to industry perspectives, possibly at an earlier point in their education. Finally, expanding the use of events such as the annual symposium sponsored by the NPS's Acquisition Research Program (ARP) may be another way to improve co-education. Once a year, the ARP hosts an Annual Acquisition Research Symposium that draws a diverse audience from academia, government, and industry, and the NPS acquisition student body attends panel sessions.[3]

Recommendations for Congress

Relax Legislative Restrictions on Backfilling Positions When Personnel Participate in Industry Rotations

Industry rotations for civilian personnel, such as the Cyber and Information Technology Exchange Program and the newly authorized Public-Private Talent Exchange, do not allow organizations to backfill positions that are temporarily vacated by participants in the programs. Interviewees indicated that this can make personnel unwilling to apply for the programs and can discourage supervisors from approving participation. In addition, as reported earlier, HCI leadership has recommended that the FY 2020 NDAA include an amendment to the Public-Private Talent Exchange authority established in Section 1104 of the FY 2017 NDAA that removes the backfill prohibition. Congress should consider accepting HCI's recommendation or, at a minimum, allowing waivers to this restriction in certain circumstances.[4]

Promote Greater Use of DAWDF

DAWDF can facilitate building business acumen, knowledge of industry operations, and industry motivation in several ways. Accordingly, we suggest that Congress protect current funding levels and, if DoD demonstrates that it can fully execute the funds

---

[3] See NPS, "Acquisition Research Program," undated(a).

[4] The authority to allow backfills would need to be exercised carefully because individuals who volunteer for a rotation may need assurance that they can return to their job after the rotation is completed.

year after year, possibly increase funding. Interviewees consistently spoke positively about the T&D gains that DAWDF enabled, which include participation in internal rotational assignments, tuition assistance for degree programs, and loan repayment. A representative from one DACM office hopes to use DAWDF to provide incentives for military and civilian AWF members to gain professional certifications. DAWDF was also used for investments in one organization's centralized personnel management system, which helped with tracking completion of DAWIA and continuous learning requirements, and DAWDF funds the AWQI. Congress should continue to provide funding for such initiatives and encourage more-expansive use of DAWDF to support further investment in the data infrastructure needed to track T&D assignments and accomplish meaningful evaluations of T&D effectiveness, external T&D in particular. DAWDF could also be used to promote industry-based rotations by helping to cover the logistical costs associated with the rotations and perhaps even to cover the cost of backfilling positions temporarily vacated when defense acquisition personnel depart for their temporary assignment in industry.

Give Actions Taken to Address AWF Knowledge Gaps Sufficient Time to Have an Effect

Developing and scaling effective solutions to closing knowledge gaps in the AWF— even those readily apparent—may take time. For example, the DCAT assessment process, a first step to establishing a knowledge standard against which to gauge gaps, is estimated to take 11 months per career field. Even efforts that are rapidly implemented, such as DAU courses and mission assistance efforts or external universities' courses (e.g., the "Better Business Deals" course), require time to yield their intended benefits, especially given the relatively low throughput of AWF personnel. In addition, the impact of an avoided or closed gap may not be obvious because some mission-related outcomes are long term; for example, an industry rotational program graduate may implement in his or her program a cost-saving practice that was acquired through that experience, but savings might not be realized or recognized immediately. That stated, deadlines to ensure that actions are taken (as available resources permit) and impacts are eventually assessed could be useful to ensure continued forward movement in meeting industry-related knowledge requirements. Moreover, while intended outcomes may not be attained in short order, monitoring changes in output-based measures in the short term (e.g., increased participation in industry rotations or increased attendance at industry-offered courses) could serve as an indicator of future gains— and could also provide evidence to inform evaluations of T&D effectiveness.

Final Thoughts

With neither standard definitions for knowledge related to business acumen, industry operations, and industry motivation, nor estimates of the necessary proficiency in each, it was difficult to address Congress's questions about gaps in those areas and the best mix of T&D to close them. Nonetheless, our research showed that DoD uses a wide variety of internal and external T&D resources to avoid those gaps and takes targeted steps to close gaps once they become apparent. Clarifying definitions, standardizing competency models, and improving approaches to gap assessment will enable DoD to better determine which career fields have a need for access to more T&D opportunities. Managing internal and external T&D resources as a portfolio and improving the evaluation of their ability to confer Section 843–related knowledge will enable better decisions about the best use of those resources for different AWF populations. Congress has helped to increase the opportunities for AWF personnel to learn from industry through means such as DAWDF and the new Public-Private Talent Exchange, but it can do more to enable DoD to take advantage of external training opportunities. Taken together, these efforts will help DoD to develop a highly skilled AWF.

Interview Methodology

From May to August 2018, we conducted 44 interviews with DoD senior leaders and SMEs and with external stakeholders.[1] A breakdown of these interviews follows:

- 26 interviews with DoD senior leaders and SMEs, including
  - directors (or acting directors) from all five of DAU's centers
  - DACMs for the three military departments and DoD's Fourth Estate (four interviews)
  - functional leaders (or their designees for interview purposes) for all the AWF career fields (12 interviews)
  - DoD HCI leadership (one interview)
  - additional DoD personnel from DAU and the military departments (four interviews)
- seven interviews with representatives of professional associations
- five interviews with representatives of three private-sector organizations with industry rotation programs or extensive in-house corporate universities
- four interviews with representatives of three universities that were providing customized courses for AWF personnel at the time of this study
- two interviews with members of the Section 809 panel, a congressionally mandated panel tasked to identify ways to improve the defense acquisition process.

Our interview sample was not a random one; instead, interviewees were intentionally chosen to cover categories of research significance. DoD interviewees were selected by virtue of their position. For example, according to DoD Instruction 5000.66, *Defense Acquisition Workforce Education, Training, Experience, and Career Development Program*, functional leaders are responsible for establishing and maintaining career-field competency models, and DACMs manage acquisition career T&D opportunities—and thus both were important interview candidates. We selected professional associations that corresponded with AWF career fields or engaged in advo-

---

[1] This study completed all necessary RAND administrative processes related to human subjects protection and received the determination that it met DoD's "not human subjects research" definition.

cacy efforts related to defense or federal acquisition, often reaching out to those mentioned in interviews. The Section 809 interviews came about via a referral from one of those associations. The universities represented in interviews were already providing executive education for DoD, which helped us gain agreement to participate and conduct the interviews within the short time frame we had. Similarly, the companies in our sample were referred to us by our DoD interviewees. For the external part of our interview sample, because some interviewees felt that they needed to obtain permission from general counsel to speak officially for their organization, we decided that all of the external interviews would be unattributed. This also helped us to avoid any implied endorsement for the associations and educators included in our sample.

Overall, we achieved the desired range of stakeholders in our sample, but we were unable to conduct as many interviews with representatives from professional associations and commercial firms as we had planned. In some cases, organizations did not reply to our requests for interviews, and others declined to participate. In addition, it is important to note that our sample is not representative, which means that we cannot estimate how prevalent these views are in organizations more generally.

We used a semi-structured approach for these interviews, which means that our interview protocol set forth opening questions and clear instructions, but we had discretion to delve into potentially fruitful lines of inquiry as they emerged. Semi-structured interviews allow the conversation between the interviewer and the participant to flow as necessary to explore issues thoroughly and permits the interviewer to curtail time spent on questions answered in earlier responses or those less relevant given the nature of the discussion. The semi-structured interview is the type of interview most frequently applied in professional contexts. Semi-structured interviews are well suited for studies that involve people accustomed to efficient use of their time, such as DoD acquisition professionals, who likely would have neither the time nor the inclination to participate in a series of free-flowing, unstructured interviews.[2] The semi-structured nature of the interviews also means that some of the questions that we posed to interviewees varied, and some comments were direct responses to our questions, while others were shared spontaneously.

Table A.1 provides a list of major topics covered in the interviews, broken down by DoD and external interviewees. In general, interview topics were aligned with assessment parameters specified in Section 843. Most notably, in all of the interviews, we asked about perceived gaps in business acumen, knowledge of industry operations, and knowledge of industry motivation for AWF personnel and extensively discussed current and potential uses of external T&D options to provide those types of knowledge. For the DoD interviews, we probed more deeply into career field–level issues and DoD

---

[2]   For more information regarding the use of semi-structured interviews—in particular, for expert or elite interviewing—see J. D. Aberbach and B. A. Rockman, "Conducting and Coding Elite Interviews," *PS Political Science and Politics*, Vol. 35, No. 4, 2002, pp. 673–676; and B. DiCicco-Bloom and B. F. Crabtree, "The Qualitative Research Interview," *Medical Education*, Vol. 40, No. 4, 2006, pp. 314–321.

Table A.1
Breakdown of Interview Topics by Interviewee Type

Topics	DoD Interviewees	External Interviewees
Section 843 knowledge type definitions (business acumen, industry operations, and industry motivation)	✓	
Perceived needs for Section 843 knowledge types	✓	✓
AWF career-field knowledge and competency requirements	✓	
T&D activities that confer the types of knowledge cited in Section 843	✓	✓
Perceived gaps related to Section 843 knowledge types	✓	✓
Role of external T&D in closing gaps	✓	✓
Facilitators of and barriers to DoD use of external T&D options	✓	
Processes related to communicating and tracking T&D activities	✓	
Processes related to evaluation of T&D	✓	✓
Recommendations to close or avoid gaps in knowledge related to business acumen, industry operations, and industry motivation	✓	✓

processes, such as those related to competency models, gaps assessments, T&D tracking, and evaluation, with questions that varied depending on the interviewee's position (e.g., we covered the same topics differently for functional leaders and DACMs).

Interviews were conducted by three members of the RAND study team. The interviews were audio-recorded, professionally transcribed, and subsequently analyzed using a computer-assisted qualitative data analysis procedure referred to as "coding." Codes are essentially tags used to organize qualitative data by topic and other characteristics.[3] The interviews were coded using QSR NVivo 12, a software package that enables its users to review, categorize, and analyze qualitative data, such as text, visual images, and audio recordings. NVivo 12 permits researchers to assign codes to passages of text and later retrieve passages of similarly coded text within and across documents. NVivo 12 is also capable of simple word-based searches and more-sophisticated text searches, such as Boolean searches involving combinations of codes.

Three members of the study team worked together to develop a coding "tree"—a set of labels for assigning units of meaning to information compiled during a study. The coding tree is provided in Box A.1 and served as the basis for a codebook that the team developed to clarify how the codes would be operationalized and to promote intercoder agreement.[4] The codebook contained code names, definitions, inclusion

[3] Miles and Huberman, 1994.

[4] DeCuir-Gunby, Marshall, and McCulloch, 2011.

Box A.1
Code Tree for 2018 RAND Section 843 Study Interviews

01 Industry Organization Background

02 Knowledge or Competency Requirements
  02.01 Career or General Requirements, Sources
  02.02 Service-Specific Differences in Competencies, Reqs
  02.03 Perceived Accuracy, Completeness of Competencies, Reqs
  02.04 Revision, Development of Competencies, Reqs
      02.04.01 Roles, Freq in Development, General Process
      02.04.02 Use of Industry Certifications or Outside Sources
      02.04.03 Use of Competency Assessments
      02.04.04 Other Development

03 Section 843–Related Knowledge Definitions
  03.01 Business Acumen
  03.02 Industry Motivation
  03.03 Industry Operations
  03.04 Other Industry Definitions
  03.05 Definition Difficulties

04 Section 843–Related Knowledge—Perceived Needs and Gaps
  04.01 Sec 843–Related Knowledge Needs
  04.02 Career Fields Most and Least Needing
  04.03 Gaps
      04.03.01 No Documented Gaps, Things Not That Bad
      04.03.02 Industry Motivation or Operations Gaps
      04.03.03 Business Acumen Gaps
      04.03.04 Negotiation Gaps
      04.03.05 Cost and Price Analysis Gaps
      04.03.06 Requirements Related Gaps
      04.03.07 Other Specific Gaps
  04.04 Especially Critical Gaps
  04.05 Past, Current Solutions to Gaps
  04.06 Gap Studies, Assessments

05 Business-Related Knowledge Sources
  05.01 Description of, Usefulness of Sources
      05.01.01 DoD—DAU
      05.01.02 DoD—Service- and Agency-Specific Schools
      05.01.03 DoD—Mobile Course Offerings
      05.01.04 DoD—Internal Rotation Assignments
      05.01.05 Non-DoD—Colleges and Universities
      05.01.06 Non-DoD—Industry
      05.01.07 Non-DoD—Professional Associations
      05.01.08 Non-DoD—Commercial Vendors
      05.01.09 Primary, Best Sources
      05.01.10 Industry Professionals
      05.01.11 OJT
  05.02 Difference in Opportunity
      05.02.01 Civilian vs Military
      05.02.02 DAWIA Certification Level, Seniority, or Rank
      05.02.03 Career Field

06 Importance of Non-DoD Options for Business-Related Knowledge
  06.01 Non-DoD Options Are Useful
  06.02 Non-DoD Options Not Needed
  06.03 Non-DoD Personnel Involvement Useful

07 Training and Development Tracking
  07.01 Did Track and How Tracked
  07.02 Not Tracked by Interviewee

08 Capacity Issues
08.01 Number of Attendees
08.02 Perceived Lack of Capacity
08.03 Sufficient Capacity

09 Training and Development Opp Communication
09.01 Means of Communication
09.02 Sufficiency or Freq of Communication

10 Barriers to Non-DoD Training and Development
10.01 Backfill Issues
10.02 Personnel Too Busy
10.03 Funding
10.04 Conflict of Interest or Propriety Data Issues

11 Facilitators of Non-DoD Training and Development
11.01 Funds
11.02 Partnerships

12 Perceived Need for Greater Use of Non-DoD Options
12.01 Need Exists
12.02 Need Does Not Exist

13 Effectiveness of Business-Related Training and Development
13.01 No Evaluation
13.02 Difficulty of Evaluation
13.03 Student Surveys and Course Evaluations
13.04 Supervisor Surveys
13.05 Financial Benefits
13.06 Other Evaluation Techniques
13.07 Instructor or Leader Surveys

14 Training and Development Option Selection

15 Messages to Congress
15.01 Positive Sentiment, Helpful
15.02 Negative Sentiment

16 Recommendations
16.01 Congress
16.02 DoD
16.03 Other

and exclusion rules, and examples of interview passages that corresponded to each code. We employed a "structural" coding approach for this study; codes were based on our study objectives and interview questions and were intended to help us identify themes.[5] Two members of the study team each applied half of the parent codes (i.e., the highest-level codes) to the full set of interviews, with one of the senior members of the study team "auditing" the coding to ensure consistent application of the codes. After the parent-level coding was completed, the study team met to develop "child" codes— a set of additional codes intended to parse out parent codes into discrete themes. The codebook was revised to include the new codes, and all three members of the coding team applied the new child codes to the parent codes. After the coding was complete, we generated coding reports that enabled us to review all of the passages tagged with a specific code together.

---

[5]   Saldaña, 2016.

The nonrandom nature of our interview sample and the semi-structured nature of our interviews suggest that it would not be appropriate either to base findings solely on interview counts or to report them. Instead, we considered strength of interview evidence when identifying themes, meaning that we considered not only how frequently a topic was coded within and across interviews but also the richness of the discussion and the level of agreement across interviewees regarding a specific topic or theme. Following Bernard,[6] we identified "exemplar quotes" —verbatim passages from the interviews—to help report readers understand themes quickly and without jargon. Such exemplar quotes are included throughout the report, and we provide the participant's organization type to help convey theme ubiquity and to show that we are not serially quoting any single individual.

---

[6]   Bernard, 2002.

Competency Model and DAU Course Analysis

Chapter Two included an overview of the methodology we used to analyze career-field competency models and DAWIA certification requirements to determine the extent to which these sources indicate that members of each career field need business acumen, knowledge of industry operations, and knowledge of industry motivation. This appendix details our methodological approach, describes the sources we drew upon in making our determinations, and provides a full list of DAU courses we categorized as conveying knowledge related to one or more of the key terms.

Overview of Process and Application of Definitions

We reviewed career field–level competency models developed by the FIPTs to determine the required knowledge sets for members of each career field related to business acumen, industry operations, and industry motivation. We also reviewed training courses and continuous learning modules (CLMs) offered by DAU to identify those that confer these three types of knowledge. In doing so, we flagged courses and modules that are required or recommended for DAWIA certifications for each level of each career field. Figure B.1, from DoDI 5000.66, illustrates the intended relationship between the career-field competency models and DAU "learning assets."

As noted throughout the report, the lack of official, codified definitions of business acumen, industry operations, and industry motivation complicates the assessment of which competencies and courses involve knowledge related to one or more of the Section 843 terms. Therefore, in making our determinations, we relied on the definitions listed in Chapter One that we developed through an analysis of our interviews. Again, these definitions are as follows:

- Business acumen: In addition to the ability to manage human, financial, and information resources strategically (OPM definition), business acumen is an understanding of industry behavior and trends that enables one to shape smart business decisions for the government.

Figure B.1
Mapping of Competency Models to DAU Learning Assets

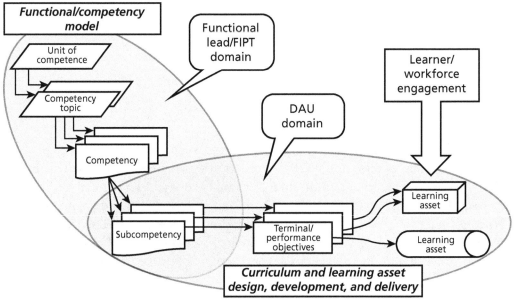

SOURCE: DoDI 5000.66.

- Industry operations: This includes plans and procedures used within an industry to provide a product or service. The need for knowledge of specific practices may vary depending on an employee's contribution to the acquisition mission. Some industry operations may be business oriented, while others may be at the confluence of business and technical knowledge—i.e., "techno-business" (e.g., milestone reviews).
- Industry motivation: This includes the range of considerations and motivations that factor into the decisionmaking of organizations in industry, including profit and revenue, market share, management and employee incentives, shareholder considerations, perspectives on risk, and the need to maintain position in a competitive environment. The relative weights of these factors may vary by industry and over time.

In some cases, it was clear that a competency or course involved knowledge related to one or more of the Section 843–related knowledge types (e.g., the competency or course included "industry" or "business acumen" in its name). In other cases, a course or competency did not explicitly mention one of these terms—but could be considered to involve knowledge related to one or more of them using the broad definitions presented above. Keywords that indicated that a competency or course involved the business-related knowledge specified in Section 843 included "industry," "business," "contractor," and "commercial." However, we erred on the side of inclusion, identifying as Section 843–related those competencies and courses that relate to industry capabilities, cost and pricing practices, and the government-contractor relationship throughout the acquisition process, as well as other types of knowledge that our interviewees mentioned in discussing their conception of the Section 843 terms.

Central to a number of the competencies and courses that we flagged is that—in order to make good business deals for the government—members of the acquisition workforce need to understand what constitutes a reasonable expectation from industry in terms of cost, schedule, and technical capability. This can be in the context of developing and negotiating a contract, overseeing and assessing contractor performance, or managing life cycle costs of systems being developed and deployed. We also identified as Section 843–related those competencies and courses that involve knowledge related to the definition of "business acumen" in OPM's executive core qualifications (ECQ) ("the ability to manage human, financial, and information resources strategically").[1] Because the components of the ECQ definition include financial, human capital, and technology management (including "keeps up-to-date on technological developments"), this resulted in flagging a number of items that relate to managing employees, technologies, and resources, even if not in the context of interacting with industry.

As this discussion underscores and as noted in Chapter Two, an overarching limitation of our competency model and DAWIA requirements analysis is that the Section 843 terms are open to interpretation. We endeavored to be as consistent as possible in applying our definitions of business acumen, industry operations, and industry motivation in determining whether competencies or courses impinged on them, but there is an unavoidable degree of imprecision involved.

Competency Model Review

We reviewed a range of career-field competency models, which varied in structure, naming conventions for the levels of competencies, and amount of detail. In some cases, we received a version of the competency model directly from the career-field functional leaders that we interviewed, and in these cases, we used the models we received directly as the basis of our knowledge needs analysis. For other career fields, we relied on the version of the competency model embedded in the AWQI e-workbook and supplemented the competencies in that workbook with those included in a workbook with "non-acquisition-unique" competencies supplied to us by an official at DAU who was involved in developing the AWQI materials.[2] Our understanding from a con-

---

[1]   OPM, undated.

[2]   The AWQI e-workbook is available at DAU, "Acquisition Workforce Qualification Initiative e-Workbook," undated(c).

versation with this individual is that these "non-acquisition-unique" competencies were not incorporated into the AWQI materials because they were difficult to translate into discrete, measurable products and tasks—but that they are part of the career-field competency models and should be included in our analysis.[3]

Table B.1 lists the career fields and describes the models we reviewed for each of them. Note, again, that when a full competency model was sent directly to us (i.e., when the "Model Received Directly" column is other than "N/A"), we based our analysis on this model rather than the combination of AWQI and the "non-acquisition-unique" workbooks. In general, however, the models received directly corresponded to the sum of the versions in AWQI and the "non-acquisition-unique" workbooks, though there were exceptions, most notably for Program Management and Science and Technology Management, both of which recently revised their models and shared these updated models with us.[4]

Notably, we found that there was overlap between competencies included in the "non-acquisition-unique" workbook (those excluded from AWQI) and those that related to the types of business-related knowledge specified in Section 843. This was especially true when it came to some of the "soft skills" that factor into the OPM definition of *business acumen* (e.g., leadership and team-building skills that are part of human capital management), but there was also an explicit "industry" category of "non-acquisition-unique" competencies that included 18 competencies across three career fields, as well as categories related to customer management, costs and finances, and suppliers. Figure B.2 is drawn from the "non-acquisition-unique" workbook and tabulates the number of competencies by category of "non-acquisition-unique" competencies and the number of career fields that had competencies in these categories.

In identifying which competencies related to one or more of the Section 843–related knowledge types, we made the determination at the subcompetency (or competency element) level, consistent with Figure B.1, which shows that it is at the subcompetency level that knowledge requirements are transmitted to DAU to guide DAU's development of learning assets to facilitate AWF members gaining the required knowledge. While there were differences across the competency models, we endeavored to identify in each model which level constituted the subcompetency level and flagged a subset of subcompetencies as Section 843–related. We also iteratively developed a tagging scheme to categorize competencies that embodied business acumen, knowledge of industry operations, and/or knowledge of industry motivation, generating category

---

[3]   Targeted discussions with DoD SME, June 21, 2018, and October 16, 2018.

[4]   In the case of Program Management, while AWQI does map to the updated model for the "acquisition-unique" competencies, the "non-acquisition-unique" workbook we reviewed was derived from an outdated version of the model; hence, summing competencies in AWQI and the other workbook did not yield the full model we received.

Table B.1
Overview of Competency Models Reviewed

Career Field	AWQI e-Workbook	"Non-Acquisition-Unique" Workbook	Model Received Directly
Business[a]	33 competencies, 89 competency elements	10 competencies that are not in AWQI	N/A
Contracting	28 competencies, 52 competency elements	10 competencies that are not in AWQI	"Contracting Competency Model.pdf": 28 "technical" competencies (the ones in AWQI), and 10 "professional" competencies (the 10 in the "non-acquisition-unique" workbook); also received "FINAL Contracting Competency Matrix with 5 anchors 2-7-08.xls": 28 competencies (the ones in AWQI), along with lists of "Examples of Supporting Knowledges" and 5 "Proficiency Level Standards" for each of the 52 competency elements included
Engineering	21 competencies, 64 competency elements	20 competencies that are not in AWQI	"ENG Career Field Competency Model.pdf": 75 competencies corresponding to 41 topics (the 41 topics correspond to the 21 competencies in AWQI plus the 20 competencies in the "non-acquisition-unique" workbook)
Facilities Engineering	26 competencies, 64 competency elements	10 competencies that are not in AWQI	N/A
Industrial Contract and Property Management	13 competencies, 53 competency elements	14 competencies that are not in AWQI	N/A
Information Technology	27 competencies, 40 competency elements	8 competencies corresponding to 13 competency elements that are not in AWQI	N/A
Life Cycle Logistics	87 competencies, 363 competency elements	N/A	N/A
Production, Quality, and Manufacturing (PQM)	18 competencies, 92 competency elements	20 competencies that are not in AWQI	"PQM Career Field Competency Model.pdf": 38 competencies corresponding to 112 subcompetencies (the 18 competencies and 92 subcompetencies that correspond to the same counts for competencies and competency elements in AWQI plus the 20 competencies and 20 subcompetencies from the "non-acquisition-unique" workbook)

Table B.1—continued

Career Field	AWQI e-Workbook	"Non-Acquisition-Unique" Workbook	Model Received Directly
Program Management	50 competencies, 153 competency elements	10 competencies corresponding to 47 competency elements that are not in AWQI	"Prog Mgmt Funct Competencies_160906.pdf": 4 umbrella categories (acquisition management, business management, technical management, and executive leadership), with a total of 18 units of competency, 70 competencies, and 190 competency elements; includes 3 levels of descriptions for each competency element
Purchasing	17 competencies, 25 competency elements	12 competencies that are not in AWQI	N/A
Science and Technology Management	6 competencies, 33 competency elements	4 competencies corresponding to 18 competency elements that are not in AWQI	"STMNewCompv2" Excel file: 69 competencies under 24 topics and 4 units of competency
Test and Evaluation	17 competencies, 46 competency elements	6 competencies corresponding to 18 competency elements that are not in AWQI	"FY 2019 TE Workforce Competency Model.pdf": 25 competencies corresponding to 69 competency elements—the 17 in AWQI plus the 6 in the "non-acquisition-unique" workbook plus an additional 2 included in a separate "Baseline for RAND Study" file[b]

[a] There was not a separate model for Business-Cost Estimating versus Business-Financial Management.

[b] We also received from DAU a "Careerfield Competency Baseline for RAND Study" workbook that included a competency model for the career fields for which we did not receive a model directly from the functional leaders; the models in this workbook were identical to those in AWQI except that they did not include the products and tasks that are layered onto the competency elements in AWQI. The only exception was Test and Evaluation, for which there were two competencies in this separate file that were not included in AWQI. Auditing was not included in AWQI or the "non-acquisition-unique" workbooks, and its competencies are structured differently enough from the remainder of the career fields to preclude a direct comparison.

tags that reflected elements of our definitions of the three key terms or types of knowledge that came up in our interviews.

In some cases, a subcompetency itself did not clearly involve knowledge related to business acumen, industry operations, or industry motivation, but the higher-level competency or unit of competency did suggest that it involved such knowledge—and, in these cases, we flagged the higher-level category as meriting inclusion. In other cases, when reviewing the competency models included in the AWQI e-workbook (the default models that we reviewed in the absence of receiving a model for a given career field directly from a functional leader), we leaned on information included in

Figure B.2
Non-Acquisition-Unique Competencies

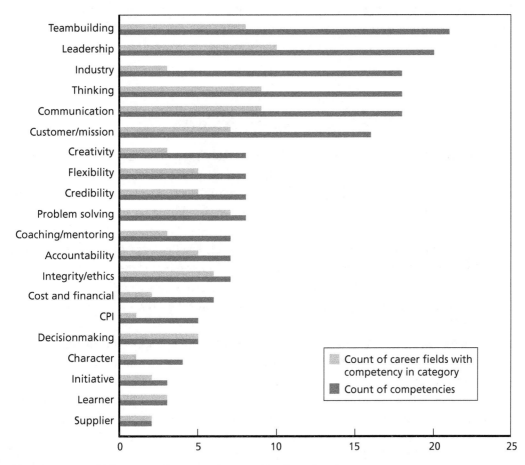

SOURCE: DAU, "AWQI NonAcq Competencies," workbook, undated.

the even-more-granular "products" and "tasks" that correspond to subcompetencies to determine whether the subcompetency itself was Section 843–related; this is consistent with guidance we received from the DoD official responsible for developing the AWQI on how it can be utilized.

We used our reviewed, tagged competency models to inform a determination of whether each career field has a comparatively lower or higher need for the types of business-related knowledge specified in Section 843, as indicated by competency model requirements. This process and its results are described in Chapter Two. We based our determinations on the breadth of types of knowledge related to business acumen, industry operations, and industry motivation included in the models rather than on the raw number of Section 843–related competency elements. This decision was rooted in the variation across the competency models we reviewed and the challenges of tabulating competency elements when the connection to the types of knowledge specified in Section 843 was most directly apparent at a higher level of aggregation (e.g., unit of competency). We recognize that this process nonetheless may favor longer, more detailed models to the extent that these models cover a broader array of topics.

DAU Course and DAWIA Requirements Review

We reviewed DAU training course and CLM names, concept cards, and course objectives (when available, which is typically the case for courses but not for modules) to identify which convey knowledge of business acumen, industry operations, and/or industry motivation, again using the definitions we developed to guide our decisions.[5] Initially, we focused our review on courses and modules included in the DAWIA Certification & Core Plus Development Guides for each level of each career field.[6] These guides list the courses and modules that are required or recommended for DAWIA certification at levels I, II, and III, and can be considered—along with the competency models themselves—an indicator of the types of knowledge required for members of each career field.

We later expanded our review to encompass the full set of DAU training courses, seeking to identify whether there were courses that conveyed the types of business-related knowledge specified in Section 843 but that were not required or recommended in the DAWIA guides for any levels of any career fields. We supplemented our list by incorporating courses in a briefing slide provided by an official at DAU entitled "DAU Core Courses Covering Understanding Industry Competencies" and those flagged in an unpublished 2013 DoD-sponsored study as being related to "business acumen" competencies (if they were not already included). Ultimately, we developed a list that contains two levels of Section 843–related courses: a narrower list that includes those validated by one of these two outside sources and those that appear most directly related to a narrower conception of the three key terms and a broader list that includes the full set of courses that appear to convey knowledge related to broader working definitions of the key terms.

For example, CON 170 ("Fundamentals of Cost and Price Analysis") is on the narrower list because it appears in one of the two outside sources, and CON 270 and 370 are also on the narrower list because they directly build upon CON 170. ACQ 415 ("Strategic Interface with Industry") does not appear on either of the two outside lists (probably because it was new at the time of this study) but is on our narrower

---

[5]   DAU, "Course Descriptions and Information," undated(d).

[6]   DAU, "DAWIA Certification and Core Plus Development Guides," undated(e).

list because it clearly includes industry-related content. ACQ 401 ("Senior Acquisition Course") and ACQ 404 ("Senior Acquisition Management Course") likewise are not on either of the outside lists but are on our narrower list because their course materials clearly include relevant content (e.g., for ACQ 404, "Given senior industry speakers [including industry students] discussions, the student participants will evaluate and discuss methods to improve industry and government engagement and mutually improved acquisition outcomes").[7] By contrast, ACQ 450 ("Leading in the Acquisition Environment") and ACQ 453 ("Leader as a Coach") are on our broader list because they relate to human capital management aspects of the OPM definition of business acumen but are less clearly connected to industry.

With regard to CLMs, we continued to limit our list to those that are required or recommended for at least one career field for at least one DAWIA certification level, as well as those listed either in the DAU briefing slide or in the unpublished 2013 DoD-sponsored study. The narrower module list is limited to those modules included on one of the two outside lists (the slide or the unpublished study).

Tables B.2 and B.3 at the end of this appendix include the results of our review of DAU training courses and CLMs, respectively. Courses and modules with an asterisk compose our narrower list of relevant learning assets. The tables include additional columns that note for which DAWIA certification levels of which career fields these courses are required or recommended. In some cases, courses may be required only for a subset of members of the career field at a given certification level or are an option to fulfill a DAWIA requirement.

As these tables make clear, there are many DAU courses and CLMs that appear to convey at least some knowledge related to one or more of the types of business-related knowledge specified in Section 843. Some of these courses and modules are required to achieve certification, others are recommended, and still others are not included among those required or recommended for certification at any level in any career field. In the cases of some 400-level courses that involve Section 843–related knowledge, this may be because they are targeted at more experienced personnel who already have their Level III certifications.

In part due to time constraints for this study, we were unable to dive deeper into the content of these courses and modules to assess the depth of instruction related to business acumen, industry operations, and industry motivation or to map between the categories we developed in our competency model analysis and the categories of knowledge conveyed in the DAU courses. Undoubtedly, some of these courses are more effective in delivering more of the business-related knowledge specified in Section 843 than others—for example, classroom-based training courses relative to online courses or the online-based, shorter CLMs. We also do not explore the implications of the

---

7   DAU, "ACQ 404 Senior Acquisition Management Course," last modified September 27, 2018b.

Table B.2
DAU Training Courses That Convey Section 843–Related Knowledge and Career Fields That Require or Recommend Them for DAWIA Certification

Course Number	Course Name	Required	Recommended (or Required for Some** or Option to Fulfill Requirement***)
ACQ 101*	Fundamentals of Systems Acquisition Management	BCE I; BFM I; CON II; ENG I; FE I; ICPM II; IT I; LCL I; PQM I; PM I; STM I; TE I	CON I**; PUR II
ACQ 202*	Intermediate Systems Acquisition, Part A	BCE II; BFM II; CON III; ENG II; ICPM III; IT II; LCL II; PQM II; PM II; STM II, TE II	CON II**
ACQ 203*	Intermediate Systems Acquisition, Part B	BCE II; BFM II; ENG II; IT II; LCL II; PQM II; PM II; TE II	CON III**; ICPM III; STM III**
ACQ 230	International Acquisition Integration	N/A	LCL III
ACQ 255	Services Acquisition Management Tools Course	N/A	N/A
ACQ 265*	Mission-Focused Services Acquisition	N/A	BCE II; CON III***; LCL III***; PM III
ACQ 315*	Understanding Industry (Business Acumen)	PM III	CON III***; LCL III***; PM II
ACQ 380	International Acquisition Management	N/A	N/A
ACQ 401*	Senior Acquisition Course	N/A	N/A
ACQ 404*	Senior Acquisition Management Course	N/A	N/A
ACQ 405*	Executive Refresher Course	N/A	LCL III
ACQ 415*	Strategic Interface with Industry	N/A	N/A
ACQ 450	Leading in the Acquisition Environment	N/A	BCE III; BFM III; CON III; ENG III; LCL III
ACQ 452	Forging Stakeholder Relationships	N/A	BCE III; BFM III; CON III; ENG III; LCL III; PM III
ACQ 453	Leader as Coach	N/A	CON III; ENG III; LCL III
BCF 110*	Fundamentals of Business Financial Management	BCE I; BFM I; PM III	ENG I; ENG II; IT I; PQM II
BCF 205*	Contractor Business Strategies	BFM II	N/A
BCF 220	Acquisition Business Management Concepts	BCE II; BFM II	ENG II; LCL II
BCF 221	Intermediate Financial Management Concepts	N/A	N/A

Table B.2—continued

Course Number	Course Name	Required	Recommended (or Required for Some** or Option to Fulfill Requirement***)
BCF 225	Acquisition Business Management Application	BCE II; BFM II	ENG II; LCL II
BCF 250	Software Cost Estimating	BCE II	N/A
BCF 275*	Applied Business Analysis Techniques	N/A	N/A
BCF 301	Business, Cost Estimating, and Financial Management Workshop	BFM III	N/A
BCF 330	Advanced Concepts in Cost Analysis	BCE III	ENG III
CME 130	Surveillance Implications of Manufacturing and Subcontractor Management	N/A	N/A
CME 230	Production Planning and Control (PP&C)	N/A	N/A
CMI 140	Multifunctional Surveillance of Prime Suppliers' Control of Subcontractors	N/A	N/A
CMQ 100	Quality Assurance Basics	N/A	N/A
CON 100*	Shaping Smart Business Arrangements	CON I; ICPM I; PUR I	N/A
CON 121*	Contract Planning	CON I; ICPM I; PM II; PUR II	BFM II***; BFM III; LCL II
CON 124*	Contract Execution	CON I; ICPM I; PM II; PUR II	BFM II***; BFM III; LCL II
CON 127*	Contract Management	CON I; ICPM I; PUR II; PM II	BFM III; LCL II***
CON 170*	Fundamentals of Cost and Price Analysis	CON I	N/A
CON 200*	Business Decisions for Contracting	CON II; ICPM II	N/A
CON 270*	Intermediate Cost and Price Analysis	CON II	N/A
CON 280	Source Selection and Administration of Service Contracts	CON II	LCL III
CON 290*	Contract Administration and Negotiation Techniques in a Supply Environment	CON II	N/A
CON 360*	Contracting for Decision Makers	CON III; ICPM III	N/A

Table B.2—continued

Course Number	Course Name	Required	Recommended (or Required for Some** or Option to Fulfill Requirement***)
CON 370*	Advanced Cost and Price Analysis	N/A	CON III***
ENG 301*	Leadership in Engineering Defense Systems	ENG III	N/A
ENG 302*	Advanced Systems Engineering	N/A	N/A
EVM 101*	Fundamentals of Earned Value Management	BCE I; BFM I; PM III	CON III; ENG I; IT II; LCL II***; PQM I; TE III
EVM 202*	Intermediate Earned Value Management	N/A	BFM II***; ENG III
IND 105	Contract Property Fundamentals	ICPM I	N/A
IND 205	Contract Government Property Management Systems and Auditing Concepts	ICPM II; ICPM III	N/A
ISA 101	Basic Information Systems Acquisition	IT I; PM II	BCE II; ENG I; LCL I; STM I; TE I
ISA 201	Intermediate Information Systems Acquisition	IT II	ENG II; LCL II; TE II
ISA 220	Risk Management Framework (RMF) for the Practitioner	N/A	IT II
ISA 301	Advanced Enterprise Information Systems Acquisition	IT III	ENG III; LCL III
ISA 320	Advanced Program Information Systems Acquisition	IT III	ENG III; LCL III; PM III
LOG 200	Product Support Strategy Development, Part A	LCL II	ENG II; IT III; PQM II; PM III
LOG 201	Product Support Strategy Development, Part B	LCL II	ENG II; PM III
LOG 235*	Performance-Based Logistics	LCL II	CON II; ENG II; PM III
LOG 340*	Life Cycle Product Support	LCL III	N/A
LOG 465*	Executive Product Support Manager's Course	N/A	LCL III**
PMT 252	Program Management Tools Course, Part I	N/A	BCE II; BFM III; ENG II; IT II; LCL II; PQM II; STM III**; TE II; TE III

Table B.2—continued

Course Number	Course Name	Required	Recommended (or Required for Some** or Option to Fulfill Requirement***)
PMT 257	Program Management Tools Course, Part II	N/A	BFM III; ENG III; IT III; LCL II; PQM II; STM III**; TE III
PMT 355	Program Management Office Course, Part A	PM III	BCE III; BFM III; ENG III; IT III; LCL III; PQM III; STM III**
PMT 360	Program Management Office Course, Part B	PM III	BCE III; BFM III; ENG III; LCL III
PMT 400	Program Manager's Skills Course	N/A	ENG III; LCL III; PM III
PMT 401*	Program Manager's Course	N/A	ENG III; PM III**
PMT 402*	Executive Program Manager's Course	N/A	PM III**
SBP 101*	Introduction to Small Business Programs, Part A	N/A	CON II
SBP 102*	Introduction to Small Business Programs, Part B	N/A	N/A
SBP 201	Intermediate Small Business Programs, Part A	N/A	N/A
SBP 202	Intermediate Small Business Programs, Part B	N/A	N/A
SBP 210*	Subcontracting	N/A	N/A
SBP 220*	Business Decisions for Small Business	N/A	N/A
STM 101	Introduction to Science and Technology Management	STM I	ENG I; IT I
STM 203	Intermediate Science and Technology Management	STM II	ENG II
STM 304	Leadership in Science and Technology Management	STM III	ENG III

* indicates that this was included in our narrower list of courses that relate to business acumen, industry operations, and/or industry motivation.
** indicates that this was required for some members of a given career field for DAWIA certification at the specified level.
*** indicates that this course is one of several that would fulfill a requirement.
NOTES: N/A indicates that this was not required or recommended for any career fields. BCE = Business-Cost Estimating, BFM = Business-Financial Management, CON = Contracting, ENG = Engineering, FE = Facilities Engineering, ICPM = Industrial Contract and Property Management, IT = Information Technology, LCL = Life Cycle Logistics, PQM = Production, Quality, and Manufacturing, PM = Program Management, PUR = Purchasing, STM = Science and Technology Management, TE = Test & Evaluation.

Table B.3
DAU Continuous Learning Modules That Convey Section 843–Related Knowledge and Career Fields That Require or Recommend Them for DAWIA Certification

Module Number	Module Name	Required	Recommended (or Required for Some** or Option to Fulfill Requirement***)
CLB 007*	Cost Analysis	PM I	CON III; IT I; LCL I; PQM I; TE II
CLB 029*	Rates	BCE III	ENG I
CLC 001*	Defense Subcontract Management	N/A	CON II; PQM I
CLC 004*	Market Research	N/A	CON I; LCL II; PUR I
CLC 007	Contract Source Selection	N/A	BCE I; BCE II; BFM II; CON II; LCL I; PQM I
CLC 008	Indirect Costs	N/A	BCE II; BFM I; CON II; ENG I
CLC 011	Contracting for the Rest of Us	LCL II	BFM II; ENG I; IT I; LCL I; PQM I; PM I; TE III
CLC 026	Performance-Based Payments Overview	N/A	CON II; LCL III
CLC 028	Past Performance Information	N/A	CON I; FE I
CLC 045*	Partnering	N/A	CON I; LCL I
CLC 047*	Contract Negotiation Techniques	N/A	CON II; IT III
CLC 051	Managing Government Property in the Possession of Contractors	CON II	LCL III
CLC 055	Competition Requirements	N/A	CON I; LCL III; PUR I
CLC 056*	Cost and Pricing Analysis	CON II	ENG I
CLC 057	Performance Based Payments and Value of Cash Flow	N/A	LCL II
CLC 058	Introduction to Contract Pricing	CON I; PUR I	N/A
CLC 104*	Analyzing Profit or Fee	N/A	BCE II; CON II; PUR II
CLC 108*	Strategic Sourcing Overview	N/A	CON II; FE III; LCL I
CLC 110*	Spend Analysis Strategies	N/A	CON II
CLC 112	Contractors Accompanying the Force	N/A	LCL I
CLC 131	Commercial Item Pricing	N/A	CON I; ENG III; PUR II
CLE 004	Introduction to Lean Enterprise Concepts	ENG I	IT I; LCL II; PQM I; PM II; TE I
CLE 007	Lean Six Sigma for Manufacturing	N/A	ENG II; IT II; LCL II; PQM II
CLE 008	Six Sigma: Concepts and Processes	N/A	ENG II; FE III; LCL III; PQM II; PM III

Table B.3—continued

Module Number	Module Name	Required	Recommended (or Required for Some** or Option to Fulfill Requirement***)
CLE 021	Technology Readiness Assessments	STM II	ENG I; IT III; PQM III; TE II
CLE 028	Market Research for Engineering and Technical Personnel	N/A	LCL II; PQM II
CLE 068*	Intellectual Property and Data Rights	ENG III; LCL II; STM I	IT II
CLE 069	Technology Transfer	STM III	N/A
CLE 076	Introduction to Agile Software Acquisition	N/A	ENG II; IT II; TE II
CLI 007	Technology Transfer and Export Control	N/A	LCL II
CLL 005	Developing a Life Cycle Sustainment Plan	LCL III	N/A
CLL 006	Public-Private Partnerships	N/A	LCL I; PM II
CLL 015*	Product Support Business Case Analysis	LCL III; TE III	BCE II; BFM III; ENG III; IT II
CLL 037*	DoD Supply Chain Fundamentals	N/A	LCL I
CLL 040*	Business Case Analysis Tools	N/A	LCL II
CLL 201	Diminishing Manufacturing Sources and Material Shortages (DMSMS) Fundamentals	N/A	LCL III; PM III
CLL 202	Diminishing Manufacturing Sources and Material Shortages (DMSMS) Executive Overview	N/A	LCL I
CLL 203	Diminishing Manufacturing Sources and Material Shortages (DMSMS) Essentials	N/A	ENG III; LCL III
CLM 005*	Industry Proposals and Communication	N/A	ENG III
CLM 014	Team Management and Leadership	STM III; TE III	BCE II; BFM III; ENG II; IT III; LCL III; PQM I
CLM 017	Risk Management	BFM II; ENG I; PQM I	FE I; IT III; LCL III; PM I; STM I; TE II
CLM 024	Contracting Overview	BFM II	BCE II; FE I; LCL I; PQM I; STM I
CLM 025	Commercial-Off-the-Shelf (COTS) Acquisition for Program Managers	N/A	PQM II; PM II
CLM 030	Common Supplier Engagement	N/A	LCL I

Table B.3—continued

Module Number	Module Name	Required	Recommended (or Required for Some** or Option to Fulfill Requirement***)
CLM 055	Program Leadership	N/A	ENG III; IT II; PQM III
CLM 059	Fundamentals of Small Business for the Acquisition Workforce	CON I; PUR I	LCL II
CLV 016*	Introduction to Earned Value Management	PM I; TE III	BCE I; CON III; ENG I; FE II; IT I; LCL I; STM II; TE II
CLV 017	Performance Measurement Baseline	N/A	BFM I; ENG II; IT II; PQM II; STM III**
HBS 302*	Negotiating for Results	N/A	N/A
HBS 305*	Negotiating for Results High Bandwidth	N/A	N/A
HBS 309	Coaching for Results	N/A	CON III
HBS 402*	Business Case Analysis	N/A	N/A
HBS 403*	Business Plan Development	N/A	N/A
HBS 405	Change Management	N/A	ICPM II
HBS 406	Coaching	N/A	CON III; ICPM II***; PQM III
HBS 409	Decision Making	N/A	PQM III; TE III
HBS 417*	Finance Essentials	N/A	N/A
HBS 424	Leading and Motivating	N/A	ICPM II***; PQM III
HBS 426*	Marketing Essentials	N/A	N/A
HBS 427	Meeting Management	N/A	PQM III; TE III
HBS 428*	Negotiating	CON II	N/A
HBS 434	Process Improvement	N/A	ICPM II; PQM II
HBS 437	Strategic Thinking	N/A	ICPM II; PQM II
HBS 440	Team Leadership	N/A	CON II
HBS 441	Team Management	N/A	CON II; PQM III; TE III

* indicates that this was included in our narrower list of modules that relate to business acumen, industry operations, and/or industry motivation.
** indicates that this was required for some members of a given career field for DAWIA certification at the specified level.
*** indicates that this module is one of several that would fulfill a requirement.
NOTES: N/A indicates that this was not required or recommended for any career fields. BCE = Business-Cost Estimating, BFM = Business-Financial Management, CON = Contracting, ENG = Engineering, FE = Facilities Engineering, ICPM = Industrial Contract and Property Management, IT = Information Technology, LCL = Life Cycle Logistics, PQM = Production, Quality, and Manufacturing, PM = Program Management, PUR = Purchasing, STM = Science and Technology Management, TE = Test & Evaluation.

differences in required and recommended courses across the three certification levels of the career fields (i.e., for personnel at different points in their careers).

Rather, to illustrate how DAWIA requirements may indicate the level of relative need for business acumen knowledge of industry operations and knowledge of industry motivation across the career fields, we limited our analysis to training courses only (excluding CLMs, which are shorter in duration and less likely to convey significant knowledge) and limited the set of courses we considered to those meeting our stricter standard for the types of knowledge specified in Section 843 (i.e., those listed on the slide provided by DAU, included in the 2013 unpublished study, or that appear to most closely relate to the key terms).[8] We then calculated how many of these training courses are required for DAWIA certification at any level for the career fields and used these counts as a gauge of whether each career field has a comparatively higher or lower need for the types of knowledge specified in Section 843 as indicated by the DAWIA requirements. The results of this analysis are presented in Chapter Two.

---

[8]   Note that our results are virtually unchanged if the full set of training courses is used as the basis for the analysis rather than just those on our narrower list. Information Technology is the only career field that would see a change in its relative need determination, shifting from the "lower" to "higher."

References

10 U.S.C. Chapter 87, *Defense Acquisition Workforce*. As of October 10, 2018:
https://www.law.cornell.edu/uscode/text/10/subtitle-A/part-II/chapter-87

10 U.S.C. Paragraph 1701a, *Management for Acquisition Workforce Excellence*.

10 U.S.C. Section 1721. As of October 12, 2018:
https://www.law.cornell.edu/uscode/text/10/1721

Aberbach, J. D., and B. A. Rockman, "Conducting and Coding Elite Interviews," *PS: Political Science and Politics*, Vol. 35, No. 4, 2002, pp. 673–676.

AcqNotes, "PBE Process: Defense Acquisition Workforce," last updated May 30, 2018. As of October 12, 2018:
http://acqnotes.com/acqnote/acquisitions/defense-acquisition-workforce

AFIT—*See* Air Force Institute of Technology.

Air Force Institute of Technology, "Course Information and Registration," undated. As of March 26, 2019:
https://www.afit.edu/LS/courseList.cfm

Air Force Institute of Technology, "AFIT Graduates Class of 248, March 2014," April 2, 2014. As of March 26, 2019:
https://www.afit.edu/news.cfm?article=588

Anderson, C., "Slowly, Steadily Measuring Impact," *Chief Learning Officer*, Vol. 12, No. 5, 2013, pp. 52–54.

Arthur, W., Jr., W. Bennett Jr., P. S. Edens, and S. T. Bell, "Effectiveness of Training in Organizations: A Meta-Analysis of Design and Evaluation Features," *Journal of Applied Psychology*, Vol. 88, No. 2, 2003, p. 234.

Association for Talent Development, *Evaluating Learning: Getting to Measurements That Matter*, Alexandria, Va., 2016a.

Association for Talent Development, *Experiential Learning for Leaders: Action Learning, On-the-Job Learning, Serious Games, and Simulations*, Alexandria, Va., 2016b.

Association for Talent Development, *The Science of Learning: Key Strategies for Designing and Delivering Training*, Alexandria, Va., 2017.

Beer, M., M. Finnstrom, and D. Schrader, *The Great Training Robbery*, Cambridge, Mass.: Harvard Business School, Working Paper 16-121, 2016.

Bernard, H. R., *Research Methods in Anthropology: Qualitative and Quantitative Methods, Third Edition*, Walnut Creek, Calif.: AltaMira Press, 2002.

Bernhard, H. B., and C. A. Ingols, "Six Lessons for the Corporate Classroom," *Harvard Business Review*, 1988.

Bersin, *Tuition Assistance Programs: Best Practices for Maximizing a Key Talent Investment*, Oakland, Calif.: Bersin by Deloitte, 2012.

Burke, L. A., and H. M. Hutchins, "Training Transfer: An Integrative Literature Review," *Human Resource Development Review*, Vol. 6, No. 3, 2007, pp. 263–296.

Burke, L. A., and H. M. Hutchins, "A Study of Best Practices in Training Transfer and Proposed Model of Transfer," *Human Resource Development Quarterly*, Vol. 19, No. 2, 2008, pp. 107–128.

Burke, R. P., and N. Spruill, *Implementation Memo to Add a Core Certification Course for the Business-Cost Estimating Career Field*, memorandum from Richard P. Burke, Deputy Director, Cost Assessment and Nancy Spruill, Director Acquisition Resources and Analysis, April 15, 2016.

Chief Information Officer, U.S. Department of Defense, "DoD CITEP Frequently Asked Questions," undated. As of October 30, 2018:
https://dodcio.defense.gov/In-the-News/Information-Technology-Exchange-Program/ITEP_FAQ/

DAU—*See* Defense Acquisition University.

DAU Directive 701, *Curricula and Program Evaluation*, January 14, 2013.

DAU Directive 708, *DAU Course Equivalency Program*, August 22, 2016.

DCPAS—*See* Defense Civilian Personnel Advisory Service.

DeCuir-Gunby, J. T., P. L. Marshall, and A. W. McCulloch, "Developing and Using a Codebook for the Analysis of Interview Data: An Example from a Professional Development Research Project," *Field Methods*, Vol. 23, No. 2, 2011, pp. 136–155.

Defense Acquisition University, "Acquisition Workforce Qualification Initiative," undated(a). As of October 22, 2018:
https://www.dau.mil/tools/awqi

Defense Acquisition University, *Agile Software Development*, undated(b). As of October 15, 2018:
https://www.dau.mil/acquipedia/Pages/ArticleDetails.aspx?aid=9980fe56-3c90-48e6-bbb4-d23bacdc890b

Defense Acquisition University, "Acquisition Workforce Qualification Initiative e-Workbook," undated(c). As of October 24, 2018:
https://www.dau.mil/tools/awqi/p/eWorkbook

Defense Acquisition University, "Course Descriptions and Information," undated(d). As of September 10, 2018:
http://icatalog.dau.mil/onlinecatalog/courses.aspx

Defense Acquisition University, "DAWIA Certification and Core Plus Development Guides," undated(e). As of September 10, 2018:
http://icatalog.dau.mil/onlinecatalog/CareerLvl.aspx

Defense Acquisition University, *Defense Acquisition University 2018 Catalog*, undated(f). As of October 15, 2018:
www.dau.mil

Defense Acquisition University, *Shaping the Future: 2017 Annual Report*, undated(g). As of October 15, 2018:
https://www.dau.mil/about/PublishingImages/Special%20Interest%20Areas/AnnualReport.pdf

Defense Acquisition University, *DAU E-Learning Asset Development Guide*, October 31, 2008.

Defense Acquisition University, *Equivalency Provider Reporting FY 17*, spreadsheet provided by DAU, July 2018a.

Defense Acquisition University, "ACQ 404 Senior Acquisition Management Course," last modified September 27, 2018b. As of September 10, 2018:
http://icatalog.dau.mil/onlinecatalog/courses.aspx?crs_id=7

Defense Civilian Personnel Advisory Service, *Memorandum for Department of Defense Civilian Employees and Supervisors in Mission Critical Occupations: Follow-up on the Defense Competency Assessment Tool, Initial Operating Capability*, October 8, 2014. As of October 15, 2018:
https://dodhrinfo.cpms.osd.mil/Directorates/HRSPAS/Strategic-Human-Capital-Management/Documents/DCAT_memo.pdf

Defense Civilian Personnel Advisory Service, *Defense Competency Assessment Tool (DCAT) Frequently Asked Questions (General)*, January 2015. As of October 15, 2018:
https://dodhrinfo.cpms.osd.mil/Directorates/HRSPAS/Strategic-Human-Capital-Management/Documents/DCAT_GeneralFAQ.pdf

DiCicco-Bloom, B., and B. F. Crabtree, "The Qualitative Research Interview," *Medical Education*, Vol. 40, No. 4, 2006, pp. 314–321.

DoD—*See* U.S. Department of Defense.

Donnithorne-Nicholls, P., "How to Engage Modern Learners," *Human Resources Magazine*, Vol. 22, No. 1, 2017, pp. 8–9.

The Eisenhower School, "Departments," undated(a). As of March 26, 2019:
http://es.ndu.edu/Departments/

The Eisenhower School, "Students," undated(b). As of March 26, 2019:
http://es.ndu.edu/People/Students/

Executive Development Associates, *Trends in Executive Development 2014: A Benchmark Report*, London: Pearson, undated.

Filipkowski, J., *Accelerating Leadership Development*, Chapel Hill, N.C.: Human Capital Institute and UNC Kenan-Flagler Business School, 2014.

Flynn, M. S., and A. Souksavatdy, *Return on Investment for the United States Navy's Training with Industry Program*, Naval Postgraduate School MBA Professional Report, June 2017.

Gallagher, James P. Jr., *An Examination of Technical Product Knowledge of Contracting Professionals at Air Force System Program Offices*, Naval Postgraduate School thesis, Monterey, Calif., December 2012.

GAO—*See* U.S. Government Accountability Office.

Gates, Susan M., Edward G. Keating, Adria D. Jewell, Lindsay Daugherty, Bryan Tysinger, Albert A. Robbert, and Ralph Masi, *The Defense Acquisition Workforce: An Analysis of Personnel Trends Relevant to Policy, 1993–2006*, Santa Monica, Calif.: RAND Corporation, TR-572-OSD, 2008. As of March 15, 2019:
https://www.rand.org/pubs/technical_reports/TR572.html

Grossman, R., and E. Salas, "The Transfer of Training: What Really Matters," *International Journal of Training and Development*, Vol. 15, No. 2, 2011, pp. 103–120.

Guest, G., K. M. MacQueen, and E. E. Namey, *Applied Thematic Analysis*, Thousand Oaks, Calif.: Sage Publications, 2011, pp. 267–268.

Gurdjian, P., T. Halbeisen, and K. Lane, "Why Leadership-Development Programs Fail," *McKinsey Quarterly*, January 2014.

Harvard Business Publishing, *Accelerate Leadership Development with Optimal Design: Six Key Principles*, 2016.

HCI—*See* Human Capital Initiatives.

Hernandez, E. R., *A Study of Benefits Resulting from the AFIT Education with Industry Program*, AFIT thesis AFIT/GSM/LSR/89S-18, September 1989.

Herring, S., "MOOCs Come of Age," *T+D*, Vol. 68, No. 1, 2014, pp. 46–49.

Hughes, A., "How Elearning Benefits Corporate Leadership Training: Bridging the Gap," *Leadership Excellence Essentials*, Vol. 35, No. 3, 2018, pp. 16–17.

Human Capital Initiatives, *Competency Assessment Overview*, presentation, undated(a), provided October 5, 2018.

Human Capital Initiatives, *Competency Management Process*, presentation, undated(b), provided by OUSD(A&S) on October 5, 2018.

Human Capital Initiatives, *07-11-18 FY17 AWF Rotational Assignments DAWDF Year-in-Review Report*, spreadsheet received from HCI, 2018a.

Human Capital Initiatives, "Workforce Metrics for FY18(Q3)," 2018b. As of October 9, 2018: http://www.hci.mil/about/workforce-metrics.html

Human Capital Initiatives (Office of the Under Secretary of Defense for Acquisition & Sustainment, Human Capital Initiatives), *Department of Defense Acquisition Workforce Development Fund 2017 Year-in-Review Report*, March 7, 2018c.

Human Capital Initiatives, *A&S Human Capital Initiatives (HCI) Updates Presented to Workforce Management Group (WMG)*, September 5, 2018d, briefing provided to RAND study team by WMG group member on October 15, 2018.

Jacot, M. T., J. Noren, and Z. L. Berge, "The Flipped Classroom in Training and Development: Fad or the Future?" *Performance Improvement*, Vol. 53, No. 9, 2014, pp. 23–28.

Kendall, Frank, Under Secretary of Defense for Acquisition, Technology and Logistics, *Better Buying Power 2.0: Continuing the Pursuit for Greater Efficiency and Productivity in Defense Spending*, memorandum for the defense acquisition workforce, November 13, 2012.

Kendall, Frank, Under Secretary of Defense for Acquisition, Technology and Logistics, *Implementation Directive for Better Buying Power 2.0—Achieving Greater Efficiency and Productivity in Defense Spending*, memorandum, April 24, 2013a.

Kendall, Frank, Under Secretary of Defense for Acquisition, Technology and Logistics, *Key Leadership Positions and Qualification Criteria*, memorandum for Secretaries of the Military Departments Component Acquisition Executives Directors of the Defense Agencies, November 8, 2013b.

Kirkpatrick, J. D., and W. K. Kirkpatrick, "The Feds Lead the Way in Making Training Evaluations More Effective—*The Value of Evaluation: Making Training Evaluations More Effective*," Alexandria, Va.: ASTD, 2012.

Kirkpatrick Partners, "The Official Site of the Kirkpatrick Model," undated. As of October 20, 2018: https://www.kirkpatrickpartners.com/Our-Philosophy/The-Kirkpatrick-Model

Lee, S. H., and J. A. Pershing, "Evaluation of Corporate Training Programs: Perspectives and Issues for Further Research," *Performance Improvement Quarterly*, Vol. 13, No. 3, 2000, pp. 244–260.

McInnis, K. J., *Defense Primer: The Department of Defense*, Congressional Research Service, 2016. As of October 12, 2018:
https://fas.org/sgp/crs/natsec/IF10543.pdf

McKinsey, *Building Organizational Capabilities: McKinsey Global Survey Results*, 2010.

Miles, M. B., and A. M. Huberman, *Qualitative Data Analysis: An Expanded Sourcebook, Second Edition*, Thousand Oaks, Calif.: Sage Publications, 1994.

NAVAIR, "NAVAIR University," undated. As of November 14, 2018:
http://www.navair.navy.mil/index.cfm?fuseaction=home.
VideoPlay&key=9C45729E-F4F7-4AED-B814-8AB1D49BEF4F

Naval Postgraduate School, "Acquisition Research Program," undated(a). As of November 14, 2018:
https://my.nps.edu/web/gsbpp/acquisition-research-program

Naval Postgraduate School, "Degree Programs," undated(b). As of November 14, 2018:
https://www.nps.edu/web/gsbpp/degree-programs

Naval Postgraduate School, "Dept. Graduation Rates: GSBPP (GB)—All Students," April 8, 2016. As of March 26, 2019:
https://my.nps.edu/documents/103399960/108002390/
Grad+Rates+-+GSBPP+by+Dept+and+Curric+2002-2015.
pdf/783987fd-18be-4dd1-b852-528984a021e1

Naval Postgraduate School, "NPS Degrees Conferred by Academic Year, Quarter and Type of Enrollment: Graduation AY 2000 to 2017," May 2018. As of March 26, 2019:
https://my.nps.edu/documents/103399960/107092937/IR+Website+Degrees+Conferred+to+2017.pdf/
c32f9342-0e90-4288-9c5c-bf0d1675ab05

NPS—*See* Naval Postgraduate School.

O'Donnell, K. W., *A Meeting of the Minds: Expanding Training and Understanding between Industry and Government*, Defense AT&L, January–February 2018.

Office of Personnel Management, "Senior Executive Service: Executive Core Qualifications," undated. As of October 22, 2018:
https://www.opm.gov/policy-data-oversight/senior-executive-service/executive-core-qualifications/

Office of Personnel Management, *A Guide to Strategically Planning Training and Measuring Results*, Washington, D.C., 2000.

Office of Personnel Management, *Training Evaluation Field Guide: Demonstrating the Value of Training at Every Level*, Washington, D.C., 2011.

Office of Personnel Management, *Executive Development Best Practices Guide*, Washington, D.C., 2012.

Office of Personnel Management, "Training and Development Policy Wiki," 2016. As of October 10, 2018:
https://www.opm.gov/wiki/training/Training-Evaluation.ashx

Office of the Under Secretary for Personnel and Readiness, "SECDEF Executive Fellows," undated. As of October 10, 2018:
https://prhome.defense.gov/Readiness/EducationTraining/SDEF/

OPM—*See* Office of Personnel Management.

Perez, S., *The ROI of Talent Development*, Chapel Hill, N.C.: University of North Carolina, Keenan-Flagler Business School, 2014.

Phillips, J. J., *In Action: Measuring Return on Investment (Vol 1)*, Alexandria, Va.: American Society for Training and Development, 1994.

Porter, C. H., J. E. Thomsen, R. T. Marlow, T. M. Geraghty, and A. J. Marcus, *Independent Study of Implementation of Defense Acquisition Workforce Improvement Efforts*, CNA, December 2016.

Professional Services Council, *Optimism Amid Diversity: The 9th Biennial Professional Services Council Acquisition Policy Survey*, July 2018. As of October 15, 2018:
https://www.grantthornton.com/-/media/content-page-files/public-sector/pdfs/surveys/2018/2018-PSC-Acquisition-Survey.ashx

PSC—*See* Professional Services Council.

Pub. L. 114–328, Div. A, Title XI, §1123, Dec. 23, 2016, 130 Stat. 2455.

Public Law 115-91, Fiscal Year 2018 National Defense Authorization Act (NDAA), December 12, 2017 (131 STAT.1480). As of October 10, 2018:
https://www.congress.gov/115/plaws/publ91/PLAW-115publ91.pdf

Rio, A., "The Future of the Corporate University," *Chief Learning Officer*, Vol. 17, No. 4, 2018, pp. 36–56.

Sable, J., "AAW Human Capital Strategic Plan: Year One," *Army AL&T Magazine, Career Development, HCSP*, September 5, 2017. As of November 15, 2018:
https://asc.army.mil/web/news-alt-ond17-aaw-human-capital-strategic-plan-year-one/

Salas, E., S. I. Tannenbaum, K. Kraiger, and K. A. Smith-Jentsch, "The Science of Training and Development in Organizations: What Matters in Practice," *Psychological Science in the Public Interest*, Vol. 13, No. 2, 2012, pp. 74–101.

Saldaña, J., *The Coding Manual for Qualitative Researchers*, Thousand Oaks, Calif.: Sage Publications, 2016.

Schwartz, M., K. A. Francis, and C. V. O'Connor, *The Department of Defense Acquisition Workforce: Background, Analysis, and Questions for Congress*, Washington, D.C.: Congressional Research Service, CRS Report R44758, July 29, 2016. As of October 30, 2018:
https://fas.org/sgp/crs/natsec/R44578.pdf]

Section 809 Panel, homepage, 2019. As of March 25, 2019:
https://section809panel.org/

Shanahan, P. M., "Engaging with Industry," memorandum from Patrick M. Shanahan, Deputy Secretary of Defense, March 2, 2018a.

Shanahan, P. M., "Public-Private Talent Exchange," memorandum from Patrick M. Shanahan, Deputy Secretary of Defense, July 19, 2018b.

Stanton, W. W., and A. D. Stanton, "Traditional and Online Learning in Executive Education: How Both Will Survive and Thrive," *Decision Sciences Journal of Innovative Education*, Vol. 15, No. 1, 2017, pp. 8–24.

Stevens Institute of Technology, *Technical Leadership Development Guidebook*, Systems Engineering Research Center, 2016.

Topno, H., "Evaluation of Training and Development: An Analysis of Various Models," *Journal of Business and Management*, Vol. 5, No. 2, 2012, pp. 16–22.

U.S. Army, "DAWDF Program," 2018. As of October 23, 2018:
https://asc.army.mil/web/career-development/dawdf-program/

U.S. Army Acquisition Center, "The Aerospace and Defense MBA—Class of 2015," 2015a. As of March 26, 2019:
https://asc.army.mil/web/wp-content/uploads/2015/09/admba-2015-class-photos.pdf

U.S. Army Acquisition Center, *Army Acquisition Education & Training (AET) Catalog 2015*, 2015b. As of March 26, 2019:
https://asc.army.mil/web/wp-content/uploads/2014/10/102014-AET-Catalog2015.pdf

U.S. Department of Defense, *Acquisition Workforce Strategic Plan: FY 2016–FY 2021*, undated. As of October 22, 2018:
http://www.hci.mil/docs/DoD_Acq_Workforce_Strat_Plan_FY16_FY21.pdf

U.S. Department of Defense, *Growing Civilian Leaders*, DoDI 1430.16, November 19, 2009. As of October 22, 2018:
http://www.esd.whs.mil/Portals/54/Documents/DD/issuances/dodi/143016p.pdf

U.S. Department of Defense, *Cybersecurity*, DoDI 8500.01, March 14, 2014.

U.S. Department of Defense, *Acquisition Workforce Strategic Plan: FY 2016–FY 2021*, 2015. As of October 22, 2018:
http://www.hci.mil/docs/DoD_Acq_Workforce_Strat_Plan_FY16_FY21.pdf

U.S. Department of Defense, *DoD Civilian Personnel Management System: Civilian Strategic Human Capital Planning (SHCP)*, DoDI 1400.25, Vol. 250, June 7, 2016a.

U.S. Department of Defense, *Fellowships, Legislative Fellowships, Internships, Scholarships, Training-with-Industry (TWI), and Grants Provided to DoD or DoD Personnel for Education and Training, USD/P&R*, DoDI 1322.06, October 12, 2016b.

U.S. Department of Defense, *Defense Acquisition Workforce Education, Training, Experience, and Career Development Program*, DoDI 5000.66, July 27, 2017a.

U.S. Department of Defense, *Report to Congress: Restructuring the Department of Defense Acquisition, Technology and Logistics Organization and Chief Management Officer Organization*, August 1, 2017b. As of October 22, 2018:
https://www.defense.gov/Portals/1/Documents/pubs/Section-901-FY-2017-NDAA-Report.pdf

U.S. Government Accountability Office, *Defense Acquisition Workforce: Actions Needed to Guide Planning Efforts and Improve Workforce Capability*, GAO-16-80, December 2015.

U.S. Government Accountability Office, *High-Risk Series: Progress on Many High-Risk Areas, While Substantial Efforts Needed on Others*, GAO-17-317, 2017.

Weinstein, M., "The Bottom Line on Leadership," *Training*, 2012, pp. 49–52.

Wentworth, D., H. Tompson, M. Vickers, A. Paradise, and M. Czarnowsky, "The Value of Evaluation: Making Training Evaluations More Effective," Alexandria, Va.: American Society for Training & Development, 2009.

CPSIA information can be obtained
at www.ICGtesting.com
Printed in the USA
BVHW051417050819
555098BV00016B/1015